Origins of life

Origins of life

Institute for Advanced Study, Princeton, New Jersey, USA
and Tarner Lecturer, 1985, Trinity College, Cambridge, England

The right of the
University of Cambridge
to print and sell
all manner of books
was granted by
Henry VIII in 1534.
The University has printed
and published continuously
since 1584.

CAMBRIDGE UNIVERSITY PRESS

Cambridge
London   New York   New Rochelle
Melbourne   Sydney

Published by the Press Syndicate of the University of Cambridge
The Pitt Building, Trumpington Street, Cambridge CB2 1RP
32 East 57th Street, New York, NY 10022, USA
10 Stamford Road, Oakleigh, Melbourne 3166, Australia

First published 1985

Printed in Great Britain by the
University Press, Cambridge

British Library cataloguing in publication data
Dyson, Freeman J.
Origins of life: Tarner lectures, given at
Cambridge University, January 1985.
1. Life – Origin
I. Title
577   QH325

Library of Congress cataloguing in publication data
Dyson, Freeman J.
Origins of life.
Bibliography: p. 78
1. Life – Origin – Addresses, essays, lectures.
I. Title.   II. Title: Tarner lectures.
QH325.D88   1985   577   85-19015

ISBN 0 521 30949 2

Contents

Preface

The delivery of these Tarner Lectures happened to coincide exactly with the two-hundredth anniversary of the first manned flight across the English Channel by Blanchard and Jeffries in January 1785. Like the intrepid balloonists, a public lecturer must carry supplies of hot air and of ballast to regulate his flight, hot air to be inserted when the text of the lecture is too short and ballast to be dropped when the text is too long. In preparing the lectures for publication I have been able to retrieve some of the dropped ballast and to vent some of the inserted air. I am glad also to have the opportunity in the published version to correct errors and to clarify obscurities, thanks to the sharp questions and criticisms which I received from the audience. I am grateful to my hosts at Trinity College for their hospitality and to my audiences for their patience.

The Tarner Lectures were established with the requirement that the lecturer speak 'on the philosophy of the sciences and the relations or want of relations between the different departments of knowledge'. I intended to ignore this requirement when I planned the lectures. I preferred to deal with concrete scientific problems rather than with philosophical generalities. I chose the origins of life as my theme because I judged the time to be ripe for a new experimental attack on the problem of origins. The main purpose of the lectures was to stimulate experiments. Nevertheless, it happens that the study of the origins of life touches many scientific disciplines and raises many philosophical questions. I therefore find myself, in spite of my pragmatic and unphilosophical intentions, fortuitously following Mr Tarner's wishes. It is impossible to speak for four hours about the origins of life without encountering some ideas which connect widely separated branches of science and other ideas which stray over the border from science to philosophy.

The lectures were addressed to a general University audience. The readers of this book are likewise expected to be educated but not expert. The same thing can be said of the author. I do not pretend to be an expert in biology. I have not read systematically through the technical literature. In my survey of experiments and ideas I make no attempt to be complete or even to be fair. I apologize in advance to all the people, living and dead, whose contributions to knowledge I shall ignore, especially to J. B. S. Haldane, Desmond Bernal, Sidney Fox, Hyman Hartman, Philip Anderson and Stuart Kauffman.

I am grateful to Martin Rees and Sydney Brenner for inviting me to a meeting with the title 'From matter to life', which was held at King's College, Cambridge, in September 1981. Biologists, chemists, physicists and mathematicians came together to talk about the origins of life, and in three days I acquired the greater part of my education as an evolutionary biologist. That meeting led me to the point of view which I am expressing in this book. I wish also to thank the Master and Fellows of Trinity College for inviting me to return to Cambridge as Tarner Lecturer in 1985. Neither King's nor Trinity is responsible for my opinions or my errors.

The first two chapters of this book are historical. Chapter 1 introduces the six characters who contributed the most to my thinking about the origins of life. Chapter 2 describes in greater detail the leading theories and the experimental background from which they arose. Chapter 3 is the most technical chapter. It describes my own contribution to the subject, a mathematical model which is intended to represent in abstract form the transition from chaos to organized metabolic activity in a population of molecules. Chapter 4 discusses some of the questions which the model leaves open and the implications of the model for the later stages of biological evolution. At the end of Chapter 4 I have included, in deference to Mr Tarner, an excursion into philosophy. My approach to the understanding of the origins of life emphasizes diversity and error-tolerance as life's salient characteristics. This approach leads me to draw analogies between the phenomena of cellular biology and the phenomena of ecology and cultural evolution. But the validity of these speculative analogies is in no way essential to our understanding of cellular biology

FREEMAN J. DYSON

Trinity College, Cambridge
February 1985

1 Illustrious predecessors

Schrödinger and Von Neumann

Forty-two years ago, in February 1943, at a bleak moment in the history of mankind, the physicist Erwin Schrödinger gave a course of lectures to a mixed audience at Trinity College, Dublin. Ireland was then, as it had been in the days of Saint Columba fourteen hundred years earlier, a refuge for scholars and a nucleus of civilization beyond the reach of invading barbarians. It was one of the few places in Europe where peaceful scientific meditation was still possible. Schrödinger proudly remarks in the published version of the lectures that they were given 'to an audience of about four hundred which did not substantially dwindle'. The lectures were published at the University Press here in Cambridge in 1944, in a little book (Schrödinger, 1944) with the title *What is Life?*

Schrödinger's book is less than a hundred pages long. It was widely read and was influential in guiding the thoughts of the young people who created the new science of molecular biology in the following decade. It is clearly and simply written, with only five references to the technical literature and less than ten equations from beginning to end. It is, incidentally, a fine piece of English prose. Although Schrödinger was exiled from his native Austria to Ireland when he was over fifty, he wrote English far more beautifully than most of his English and American contemporaries. He reveals his European background only in the epigraphs which introduce his chapters: three are from Goethe, in German, three are from Descartes and Spinoza, in Latin, and one is from Unamuno, in Spanish. As a sample of his style I quote the opening sentences of his preface:

A scientist is supposed to have a complete and thorough knowledge, at first hand, of *some* subjects, and therefore

I

is usually expected not to write on any topic of which he is not a master. This is regarded as a matter of *noblesse oblige*. For the present purpose I beg to renounce the *noblesse*, if any, and to be freed of the ensuing obligation. My excuse is as follows. We have inherited from our forefathers the keen longing for unified, all-embracing knowledge. The very name given to the highest institutions of learning reminds us that from antiquity and throughout many centuries the *universal* aspect has been the only one to be given full credit. But the spread, both in width and depth, of the multifarious branches of knowledge during the last hundred odd years has confronted us with a queer dilemma. We feel clearly that we are only now beginning to acquire reliable material for welding together the sum-total of what is known into a whole; but, on the other hand, it has become next to impossible for a single mind fully to command more than a small specialized portion of it. I can see no other escape from this dilemma (lest our true aim be lost for ever) than that some of us should venture to embark on a synthesis of facts and theories, albeit with second-hand and incomplete knowledge of some of them, and at the risk of making fools of themselves. So much for my apology.

This apology for a physicist venturing into biology will serve for me as well as for Schrödinger, although in my case the risk of the physicist making a fool of himself may be somewhat greater.

Schrödinger's book was seminal because he knew how to ask the right questions. The basic questions which Schrödinger asked were the following: What is the physical structure of the molecules which are duplicated when chromosomes divide? How is the process of duplication to be understood? How do these molecules retain their individuality from generation to generation? How do they succeed in controlling the metabolism of cells? How do they create the organization that is visible in the structure and function of higher organisms? He did not answer these questions, but by asking them he set biology moving along the path which led to the epoch-making discoveries of the subsequent forty years: to the discovery of the double helix and the triplet code, to the precise analysis and wholesale synthesis of genes, and to the quantitative measurement of the evolutionary divergence of species.

Schrödinger showed wisdom not only in the questions that he asked but also in the questions that he did not ask. He did not ask any questions about the origin of life. He understood that the time was ripe in 1943 for a fundamental understanding of the physical basis of life. He also understood that the time was not then ripe for any fundamental understanding of life's origin. Until the basic chemistry of living processes was clarified, one could not ask meaningful questions about the possibility of spontaneous generation of these processes in a pre-biotic environment. He wisely left the question of origins to a later generation.

Now, forty years later, the time is ripe to ask the questions which Schrödinger avoided. The questions of origin are now becoming experimentally accessible, just as the questions of structure were becoming experimentally accessible in the 1940s. Schrödinger asked the right questions about structure because his thoughts were based on the experimental discoveries of his friend Max Delbrück. We can hope to ask the right questions about origins today because our thoughts are guided by the experimental discoveries of Manfred Eigen and Leslie Orgel. Delbrück became the chief explorer of the problems of structure because he hit on the bacteriophage as the ideal experimental tool, a biological system stripped of inessential complications and reduced to an almost bare genetic apparatus. The bacteriophage was for biology what the hydrogen atom was for physics. In a similar way Eigen became the chief explorer of the problems of the origin of life in the 1970s because he hit on ribonucleic acid (RNA) as the ideal experimental tool for studies of molecular evolution in the test-tube. Eigen's RNA experiments have carried Delbrück's bacteriophage experiments one step further: Eigen stripped the genetic apparatus completely naked, thereby enabling us to study its replication unencumbered

3

by the baggage of structural molecules which even so rudimentary a creature as a bacteriophage carries with it.

Before discussing the experiments of Eigen and Orgel in detail, I want to finish my argument with Schrödinger. At the risk, again, of making a fool of myself, I shall venture to say that in his discussion of the nature of life Schrödinger missed an essential point. And I feel that the same point was also missed by Manfred Eigen in his discussion of the origin of life. I hasten to add that in disagreeing with Schrödinger and Eigen I am not disputing the greatness of their contributions to biology. I am not saying that they were, or are, wrong. I am saying only that they did not ask all of the important questions.

In Schrödinger's book we find four chapters describing in lucid detail the phenomenon of biological replication and a single chapter describing less lucidly the phenomenon of metabolism. Schrödinger finds a conceptual basis in physics both for exact replication and for metabolism. Replication is explained by the quantum-mechanical stability of molecular structures, while metabolism is explained by the ability of a living cell to extract negative entropy from its surroundings in accordance with the laws of thermodynamics. Schrödinger was evidently more interested in replication than in metabolism. There are two obvious reasons for his bias. Firstly, he was, after all, one of the inventors of quantum mechanics, and it was natural for him to be primarily concerned with the biological implications of his own brain-child. Secondly, his thinking was based on Delbrück's experiments, and these were biassed in the same direction. Delbrück's experimental system, the bacteriophage, is a purely parasitic creature in which the metabolic function has been lost and only the replicative function survives. It was indeed precisely this concentration of attention upon a rudimentary and highly specialized form of life that enabled Delbrück to do experiments exploring the physical basis of biological replication. It was necessary to find a creature without metabolism in order to isolate experimentally the phenomena of replication. Delbrück penetrated more deeply than his contemporaries into the mechanics of replication because he was not distracted by the problems of metabolism. Schrödinger saw the world of biology through Delbrück's eyes. It is not surprising that Schrödinger's view of what constitutes a living organism resembles a bacteriophage more than it resembles a bacterium or a fruit-fly. His single chapter devoted to the metabolic aspect of life appears to be an afterthought, put in for the sake of completeness but not affecting the main line of his argument.

The main line of Schrödinger's argument, which leads from the facts of biological replication to the quantum-mechanical structure of the gene, was brilliantly right and fruitful. It set the style for the subsequent development of molecular biology. Neither Schrödinger himself nor the biologists who followed his lead appear to have been disturbed by the logical gap between his main argument and his discussion of metabolism. Looking back on his 1943 lectures now with the benefit of forty years of hindsight, we may wonder why he did not ask some fundamental questions which the gap might have suggested to him: Is life one thing or two things? Is there a logical connection between metabolism and replication? Can we imagine metabolic life without replication, or replicative life without metabolism? These questions were not asked because Schrödinger and his successors took it for granted that the replicative aspect of life was primary, the metabolic aspect secondary. As their understanding of replication became more and more triumphantly complete, their lack of understanding of metabolism was pushed into the background. In popular accounts of molecular biology as it is now taught to school-children, life and replication have become practically synonymous. In modern discussions of the origin of life it is often taken for granted that the origin of life is the same thing as the origin of replication. Manfred Eigen's view is an extreme example of this tendency. Eigen chose RNA as the working material for his experiments because it is ideal for studying replication but is incapable of metabolism. Eigen's theories about the origin of life are in fact theories about the origin of replication.

5

It is important here to make a sharp distinction between replication and reproduction. I am suggesting as a hypothesis that the earliest living creatures were able to reproduce but not to replicate. What does this mean? For a cell, to reproduce means simply to divide into two cells with the daughter cells inheriting approximately equal shares of the cellular constituents. For a molecule, to replicate means to construct a precise copy of itself by a specific chemical process. Cells can reproduce but only molecules can replicate. In modern times, reproduction of cells is always accompanied by replication of molecules, but this need not always have been so in the past. Even in modern times, replication of molecules often occurs without reproduction of cells.

Only five years after Schrödinger gave his lectures in Dublin, the logical relations between replication and metabolism were clarified by the mathematician John Von Neumann (Von Neumann, 1948). Von Neumann described an analogy between the functioning of living organisms and the functioning of mechanical automata. His automata were an outgrowth of his thinking about electronic computers. A Von Neumann automaton had two essential components; later on, when his ideas were taken over by the computer industry, these were given the names hardware and software. Hardware processes information; software embodies information. These two components have their exact analogues in living cells; protein is hardware and nucleic acid is software. Protein is the essential component for metabolism. Nucleic acid is the essential component for replication. Von Neumann described precisely, in abstract terms, the logical connections between the components. For a complete self-reproducing automaton, both components are essential. Yet there is an important sense in which hardware comes logically prior to software. An automaton composed of hardware without software can exist and maintain its own metabolism. It can live independently for as long as it finds food to eat or numbers to crunch. An automaton composed of software without hardware must be an obligatory parasite. It can function only in a world already containing other

6

automata whose hardware it can borrow. It can replicate itself only if it succeeds in finding a cooperative host automaton, just as a bacteriophage can replicate only if it succeeds in finding a cooperative bacterium.

Let me summarize the drift of my argument up to this point. Our illustrious predecessor Erwin Schrödinger gave his book the title *What is Life?* but neglected to ask whether the two basic functions of life, metabolism and replication, are separable or inseparable. Our illustrious predecessor John Von Neumann, using the computer as a metaphor, raised the question that Schrödinger had missed and gave it a provisional answer. Von Neumann observed that metabolism and replication, however intricately they may be linked in the biological world as it now exists, are logically separable. It is logically possible to postulate organisms that are composed of pure hardware and capable of metabolism but incapable of replication. It is also possible to postulate organisms that are composed of pure software and capable of replication but incapable of metabolism. And if the functions of life are separated in this fashion, it is to be expected that the latter type of organism will become an obligatory parasite upon the former. This logical analysis of the functions of life helps to explain and to correct the bias toward replication that is evident in Schrödinger's thinking and in the whole history of molecular biology. Organisms specializing in replication tend to be parasites, and molecular biologists prefer parasites for experimental study because parasites are structurally simpler than their hosts and better suited to quantitative manipulation. In the balance of nature there must be an opposite bias. Hosts must exist before there can be parasites. The survival of hosts is a precondition for the survival of parasites. Somebody must eat and grow to provide a home for those who only reproduce. In the world of microbiology, as in the world of human society and economics, we cannot all be parasites.

When we begin to think about the origins of life we meet again the question which Schrödinger did not ask: What do we mean by life? And we meet again Von Neumann's answer, that life is not one thing but two, metabolism and replication, and that the two things are logically separable. There are accordingly two logical possibilities for life's origins. Either life began only once, with the functions of replication and metabolism already present in rudimentary form and linked together from the beginning, or life began twice, with two separate kinds of creatures, one kind capable of metabolism without exact replication, the other kind capable of replication without metabolism. If life began twice, the first beginning must have been with proteins, the second beginning with nucleic acids. The first protein creatures might have existed independently for a long time, eating and growing and gradually evolving a more and more efficient metabolic apparatus. The nucleic acid creatures must have been obligatory parasites from the start, preying upon the protein creatures and using the products of protein metabolism to achieve their own replication.

The main theme of these lectures will be a critical examination of the second possibility, the possibility that life began twice. I call this possibility the double-origin hypothesis. It is a hypothesis, not a theory. A theory of the origin of life should describe in some detail a postulated sequence of events. The hypothesis of dual origin is compatible with many theories. It may be useful to examine the consequences of the hypothesis without committing ourselves to any particular theory.

I do not claim that the double-origin hypothesis is true, or that it is supported by any experimental evidence. Indeed my purpose is just the opposite. I would like to stimulate experimental chemists and biologists and paleontologists to find the evidence by which the hypothesis might be tested. If it can be tested and proved wrong, it will have served its purpose. We will then have a firmer foundation of fact on which to build theories of single origin. If the double-origin hypothesis can be tested and not proved wrong, we can proceed with greater confidence to build theories of double origin. The hypothesis is useful only insofar as it may suggest new experiments.

Lacking new experiments, we have no justification for believing strongly in either the single-origin or the double-origin hypothesis. I have to confess my own bias in favour of double-origin. But my bias is based only on general philosophical preconceptions, and I am well aware that the history of science is strewn with the corpses of dead theories which were in their time supported by the prevailing philosophical viewpoints. For what it is worth, I may state my philosophical bias as follows. The most striking fact which we have learned about life as it now exists is the ubiquity of dual structure, the division of every organism into hardware and software components, into protein and nucleic acid. I consider dual structure to be *prima facie* evidence of dual origin. If we admit that the spontaneous emergence of protein structure and of nucleic acid structure out of molecular chaos are both unlikely, it is easier to imagine two unlikely events occurring separately over a long period of time than to imagine two unlikely events occurring simultaneously. Needless to say, vague arguments of this sort, invoking probabilities which we are unable to calculate quantitatively, cannot be conclusive. The main reason why I am hopeful for progress in the understanding of the origin of life is that the subject is moving away from the realm of philosophical speculation and into the realm of experimental science.

Eigen and Orgel

The third and fourth names on my list of illustrious predecessors are those of Manfred Eigen and Leslie Orgel. Unlike Schrödinger and Von Neumann, they are experimenters. They are the chief explorers of the experimental approaches to the problem of the origin of life. They are, after all, chemists, and this is a job for chemists. Eigen and his colleagues in Germany have done experiments which show us biological organization originating spontaneously and evolving in a test-tube (Fig. 1). More precisely, they have demonstrated that a solution of nucleotide monomers will, under suitable conditions, give rise to a nucleic acid polymer molecule which replicates and mutates and competes with its progeny for survival. From a certain point of view, one might claim that these experiments have already achieved

9

Fig. I. The Biebricher-Eigen-Luce experiment demonstrating evolution of RNA molecules in a test-tube containing a solution of nucleotide monomers with added replicase enzyme. The four curves on the left were obtained with 10^{14}, 10^6, 10^3 and 1 molecules of RNA template added to the mixture. The three curves on the right are three separate runs with no template added (Data from Eigen *et al.*, 1981).

the spontaneous generation of life from non-life. They bring us at least to the point where we can ask and answer questions about the ability of nucleic acids to synthesize and organize themselves (Eigen *et al.*, 1981). Unfortunately, the conditions in Eigen's test-tubes are not really pre-biotic. To make his experiments work, Eigen put into the test-tubes a polymerase enzyme, a protein catalyst extracted from a living bacteriophage. The synthesis and replication of the nucleic acid is dependent on the structural guidance provided by the enzyme. We are still far from an experimental demonstration of the appearance of biological order without the help of a biologically-derived precursor. Nevertheless, Eigen has provided the tools with which we may begin to attack the problem of origins. He has brought the origin of life out of the domain of idle speculation and into the domain of experiment.

Leslie Orgel, like Manfred Eigen, is an experimental chemist. He taught me most of what I know about the chemical antecedents of life. He has done experiments complementary to those of Eigen. Eigen was able to make RNA grow out of nucleotide monomers without having any RNA template for the monomers to copy, but with a polymerase enzyme to tell the monomers what to do. Orgel has done equally important experiments in the opposite direction. Orgel demonstrated that nucleotide monomers will, under certain conditions, polymerize to form RNA if they are given an RNA template to copy without any polymerase enzyme. Orgel found that zinc ions in the solution are a good catalyst for the RNA synthesis. It may not be entirely coincidental that many modern biological enzymes have zinc ions in their active sites. To summarize, Eigen made RNA using an enzyme but no template and Orgel made RNA using a template but no enzyme. In living cells RNA is made using both templates and enzymes. If we suppose that RNA was the original molecule of life, then to understand the origin of life we have to make RNA using neither a template nor an enzyme. Neither Eigen nor Orgel has come close to achieving this goal.

The experiments of Eigen and Orgel fit more naturally into the framework of a double-origin hypothesis. According to the double-origin hypothesis, RNA was not the original molecule of life. In this hypothesis the original molecules of life were proteins, and life of a sort was already established before RNA came into the picture. In this context the Eigen and Orgel experiments are exploring the evolution of RNA under conditions appropriate to the second origin of life. They come close to describing a parasitic development of RNA-life within an environment created by a pre-existing protein-life. Concerning the first origin of life, the origin of protein-life and of protein metabolism, they say nothing. The origin of metabolism is the next great virgin territory which is waiting for experimental chemists to explore.

Margulis

The fifth name on my list of illustrious predecessors is that of Lynn Margulis. Although she is still very much alive and considerably younger than I am, she set the style in which

I came to think about early evolution. The experiments of Eigen and Orgel gave us two solid foundations of knowledge with a wide river of ignorance running between them. Since we have solid ground on the two sides, it is not hopeless to think of building a bridge over the river. In science a bridge is a theory. When bridges are to be built, theoretical scientists may have a useful role to play.

Lynn Margulis is one of the chief bridge-builders in modern biology. She built a bridge between the facts of cellular anatomy and the facts of molecular genetics. Her bridge was the idea that parasitism and symbiosis were the driving forces in the evolution of cellular complexity. She did not invent this idea, but she was its most active promoter and systematizer. She collected the evidence to support her view that the main internal structures of eucaryotic cells did not originate within the cells but are descended from independent living creatures which invaded the cells from outside like carriers of an infectious disease (Margulis, 1970, 1981). The invading creatures and their hosts then gradually evolved into a relationship of mutual dependence so that the erstwhile disease organism became by degrees a chronic parasite, a symbiotic partner and finally an indispensable part of the substance of the host. This Margulis picture of early cellular evolution now has incontrovertible experimental support. The molecular structures of chloroplasts and mitochondria are found to be related more closely to alien bacteria than to the cells in which they have been incorporated for one or two thousand million years. But there are also general philosophical reasons for believing that the Margulis picture will be valid even in cases where it cannot be experimentally demonstrated. A living cell, in order to survive, must be intensely conservative. It must have a finely tuned molecular organization and it must have efficient mechanisms for destroying promptly any molecules which depart from the overall plan. Any new structure arising within this environment must be an insult to the integrity of the cell. Almost by definition a new structure will be a disease which the cell will do its best to resist. It is possible to imagine new structures arising internally within the cell

12

and escaping its control, like a cancer growing in a higher organism. But it is much easier to imagine new structures coming in from the outside like infectious bacteria, already prepared by the rigours of independent living to defend themselves against the cell's efforts to destroy them.

The main reason why I find the two-origin hypothesis philosophically congenial is that it fits well into the general picture of evolution portrayed by Margulis. According to Margulis, most of the big steps in cellular evolution were caused by parasites. The double-origin hypothesis implies that nucleic acids were the oldest and most successful cellular parasites. It extends the scope of the Margulis picture of evolution to include not only eucaryotic cells but procaryotic cells as well. It proposes that the original living creatures were cells with a metabolic apparatus directed by enzymes (proteins) but with no genetic apparatus. Such cells would lack the capacity for exact replication but could grow and divide and reproduce themselves in an approximate statistical fashion. They might have continued to exist for millions of years, gradually diversifying and refining their metabolic pathways. Amongst other things, they discovered how to synthesize ATP, adenosine triphosphate, the magic molecule which serves as the principal energy-carrying intermediate in all modern cells. Cells carrying ATP were able to function more efficiently and prevailed in the Darwinian struggle for existence. In time it happened that cells were full of ATP and other related molecules such as AMP, adenosine monophosphate.

Now we observe the strange fact that the two molecules ATP and AMP, which have almost identical chemical structures (Fig. 2), have totally different but equally essential functions in modern cells. ATP is the universal energy carrier. AMP is one of the nucleotides which make up RNA and function as bits of information in that genetic apparatus. To get from ATP to AMP, all you have to do is remove two phosphate moieties. I am proposing that the primitive cells had no genetic apparatus but were saturated with molecules like AMP as a result of the energy-carrying function of ATP. This was a dangerously explosive situation, and in one cell

Adenosine
triphosphate

$$NH_2$$

$$-O-\overset{\overset{\displaystyle O}{\|}}{\underset{\underset{\displaystyle O^-}{|}}{P}}-O-\overset{\overset{\displaystyle O}{\|}}{\underset{\underset{\displaystyle O^-}{|}}{P}}-O-\overset{\overset{\displaystyle O}{\|}}{\underset{\underset{\displaystyle O^-}{|}}{P}}-O-CH_2-CH$$

Adenosine
5′-monophosphate

$$NH_2$$

$$-O-\overset{\overset{\displaystyle O}{\|}}{\underset{\underset{\displaystyle O^-}{|}}{P}}-O-CH_2-CH$$

Fig. 2. The molecular structures of adenosine triphosphate (ATP) and adenosine 5′-monophosphate (AMP), otherwise known as adenine nucleotide.

which happened to be carrying an unusually rich supply of nucleotides, an accident occurred. The nucleotides began doing the Eigen experiment on RNA synthesis three thousand million years before it was done by Eigen. Within the cell, with some help from pre-existing enzymes, the nucleotides produced an RNA molecule which then continued to replicate itself. In this way RNA first appeared as a parasitic disease within the cell. The first cells in which the RNA disease occurred probably became sick and died. But then, according to the Margulis scheme, some of the infected cells learned how to survive the infection. The protein-based life learned to tolerate the RNA-based life. The parasite became a symbiont. And then, very slowly over millions of years, the protein-based life learned to make use of the capacity for exact replication which the chemical structure of RNA provided. The primal symbiosis of protein-based life and parasitic RNA grew gradually into a harmonious unity, the modern genetic apparatus.

This view of RNA as the oldest and most incurable of our parasitic diseases is only a poetic fancy, not yet a serious scientific theory. Still, it is attractive to me for several reasons. Firstly, it is in accordance with our human experience that hardware should come before software. The modern cell is like a computer-controlled chemical factory in which the proteins are the hardware and the nucleic acids are the software. In the evolution of machines and computers, we always developed the hardware first before we began to think about software. I find it reasonable that natural evolution should have followed the same pattern. A second argument in favour of the parasite theory of RNA comes from the chemistry of amino-acids and nucleotides. It is easy to synthesize amino-acids, the constituent parts of proteins, out of plausible pre-biotic materials such as water, methane and ammonia. The synthesis of amino-acids from a hypothetical reducing atmosphere was demonstrated in the classic experiment of Miller in 1953. The nucleotides which make up nucleic acids are much more difficult to synthesize. Nucleotide bases such as adenine and guanine have been synthesized by Oró from ammonia and hydrocyanic acid. But to go from a base to a complete nucleotide is a more delicate matter. Furthermore, once formed, nucleotides are less stable than amino-acids. Because of the details of the chemistry, it is much easier to imagine a pond on the pre-biotic earth becoming a rich soup of amino-acids than to imagine a pond becoming a rich soup of nucleotides. Nucleotides would have had a better chance to polymerize if they originated in biological processes inside already-existing cells. My third reason for liking the parasite theory of RNA is that it may be experimentally testable. If the theory is true, living cells may have existed for a very long time before becoming infected with nucleic acids. There exist microfossils, that is traces of primitive cells, in rocks which are more than three thousand million years old. It is possible

15

that some of these microfossils might come from cells older than the origin of RNA. It is possible that the microfossils may still carry evidence of the chemical nature of the ancient cells. For example, if the microfossils were found to preserve in their mineral constituents significant quantities of phosphorus, this would be strong evidence that the ancient cells already possessed something resembling a modern genetic apparatus. So far as I know, no such evidence has been found. I do not know whether the processes of fossilization would be likely to leave chemical traces of nucleic acids intact. So long as this possibility exists, we have the opportunity to test the hypothesis of a late origin of RNA by direct observation.

Kimura

The last of the illustrious predecessors on my list is the geneticist Motoo Kimura. Kimura developed the mathematical basis for a statistical treatment of molecular evolution (Kimura, 1970) and he has been the chief advocate of the neutral theory of evolution (Kimura, 1983). The neutral theory says that, through the history of life from beginning to end, random statistical fluctuations have been more important than Darwinian selection in causing species to evolve. Evolution by random statistical fluctuation is called genetic drift. Kimura maintains that genetic drift drives evolution more powerfully than natural selection. I am indebted to Kimura in two separate ways. Firstly, I use Kimura's mathematics as a tool for calculating the behaviour of molecular populations. The mathematics is correct and useful, whether you believe in the neutral theory of evolution or not. Secondly, I find the neutral theory helpful even though I do not accept it as dogma. In my opinion, Kimura has overstated his case, but still his picture of evolution may sometimes be right. Genetic drift and natural selection are both important, and there are times and places where one or the other may be dominant. In particular, I find it reasonable to suppose that genetic drift was dominant in the very earliest phase of biological evolution, before the mechanisms of heredity had become exact. Even if the neutral theory is not true in general, it may be a useful approximation to make in building models of pre-biotic evolution.

We know almost nothing about the origin of life. We do not even know whether the origin was gradual or sudden. It might have been a process of slow growth stretched out over millions of years or it might have been a single molecular event that happened in a fraction of a second. As a rule, natural selection is more important over long periods of time and genetic drift is more important over short periods. If you think of the origin of life as being slow, you must think of it as a Darwinian process driven by natural selection. If you think of it as being quick, then the Kimura picture of evolution by statistical fluctuation without selection is appropriate. In reality the origin of life must have been a complicated process, with incidents of rapid change separated by long periods of slow adaptation. A complete description needs to take into account both drift and selection. In my calculations I have made use of the theorist's privilege to simplify and idealize a natural process. I have considered the origin of life to be an isolated event occurring on a rapid time-scale. In this hypothetical context, it is consistent to examine the consequences of genetic drift acting alone. Darwinian selection will begin its work after the process of genetic drift has given it something to work on.

If one wishes as I do to examine seriously the double-origin hypothesis, the hypothesis that life began and flourished without the benefit of exact replication, then it is natural to imagine that genetic drift remained strong and natural selection remained relatively weak during the early exploratory phases of evolution. But this is not to say that Darwinian selection had to wait until life learned to replicate exactly. Darwinian selection is not logically dependent on exact replication. Indeed, Darwin himself knew nothing of exact replication when he invoked the idea of natural selection. Darwinian selection would have operated to guide the evolution of living creatures even at a time when those creatures may have lacked anything resembling a modern genetic apparatus. All that is necessary for natural selection to operate is that there be some inheritance of chemical constituents from an organism to its progeny. The inheritance need not be exact. It is sufficient if a cell splitting into two daughter cells transmits to each of its daughters with a high probability a population of molecules capable of continuing its own pattern of metabolism. Statistical inheritance, as Darwin well knew, can be good enough. Darwinian selection is unavoidable as soon as inheritance begins, no matter how sloppy the mechanism of inheritance may be. When I apply Kimura's mathematics of genetic drift to describe the earliest phase of the first origin of life, this does not mean that I am following Kimura in his belief that genetic drift continued to be dominant even in the period between the origin of metabolism and the origin of replication.

There are many other illustrious predecessors besides those whom I have mentioned. I chose to speak of these six, Schrödinger, Von Neumann, Eigen, Orgel, Margulis and Kimura, because each of them is in some sense a philosopher as well as a scientist. Each of them brought to biology not just technical skills and knowledge but a personal philosophical viewpoint extending beyond biology over the whole of science. From all of them I have borrowed the ideas which fitted together to form my own philosophical viewpoint. The origin of life is one of the few scientific problems which is broad enough to make use of ideas from almost all scientific disciplines. Schrödinger brought to it ideas from physics, Von Neumann ideas from mathematical logic, Eigen and Orgel ideas from chemistry, Margulis ideas from ecology and Kimura ideas from population biology. What I am trying to do in these lectures is to explore the connections, to see whether mathematical logic and population biology may have raised new questions which chemistry may be able to answer.

2 Experiments and theories

The study of pre-biotic evolution divides itself into three main stages which one may label geophysical, chemical and biological. The geophysical stage concerns itself with the early history of the earth and especially with the nature of the earth's primitive atmosphere. The chemical stage concerns itself with the synthesis, by natural processes operating within plausible models of the primitive atmosphere and ocean, of the chemical building-blocks of life. By building-blocks we mean principally the amino-acids and nucleotide monomers out of which proteins and nucleic acids, respectively, are built. The biological stage concerns itself with the appearance of biological organization, with the building of a coordinated population of protein or nucleic acid polymers out of a random assortment of amino-acids or nucleotide monomers.

Generally speaking, it can be said that the geophysical and chemical stages of pre-biotic evolution are reasonably well understood. At least these two stages are in the hands of competent experts, and I have nothing significant to add to what the experts have reported. Theories of the geophysical stage can be checked by abundant observations in the fields of geochemistry, meteorite chemistry and radio-astronomy. Theories of the chemical stage can be checked by experiments done by chemists in the laboratory. Many details remain to be elucidated, but the geophysical and chemical stages are no longer shrouded in mystery. I have therefore concentrated my attention on the biological stage. The problem of the origin of life is for me the biological stage, the problem of the appearance of biological organization out of molecular chaos. It is in this biological stage that mysteries still remain. The purpose of my own work has been to try to define precisely what we mean by the appearance of biological organization and thereby to make the biological stage accessible to experimental study.

Geophysics and chemistry

Let me now return briefly to the geophysical stage. The fundamental fact which we have learned in the last twenty years from the observations of radio-astronomers using millimetre-wave telescopes is that our galaxy is thickly populated with molecular clouds containing large quantities of molecular hydrogen, water, ammonia, carbon monoxide, methyl alcohol, hydrocyanic acid and other molecules which are chemically reducing rather than oxidizing. Observation also shows that these molecular clouds are the places where stars are at present being formed by gravitational condensation of the molecule-bearing gas. It is reasonable to imagine that the earth and the sun were formed by condensation of a similar molecular cloud four-and-a-half thousand million years ago. It then follows that the primitive earth almost certainly possessed an atmosphere composed of the same volatile reducing molecules that we see in the molecular clouds today. The original atmosphere of the earth might have been stripped off many times by intense radiation from the sun or by the violent infall of planetesimals as the earth accumulated. Nevertheless, the atmosphere which remained after the solar system emerged into quiescence from the initial violence is likely still to have been mainly composed of the molecules which exist so abundantly in the sky. In the carbonaceous meteorites which were formed at roughly the same time as the earth we find large quantities of reduced carbon; this provides supporting evidence for reducing conditions in the early solar system. How long the earth's reducing atmosphere may have lasted we do not know. Geochemical evidence suggests that it was replaced by an oxidizing atmosphere about two thousand million years ago, after life was well established and photosynthetic organisms had begun to produce free oxygen in large quantities. We can reasonably postulate that at the epoch of life's origin the earth's atmosphere was reducing and contained some or all of the hydrogen-rich species water, ammonia, methane and molecular hydrogen. Other common species which are neither reducing nor oxidizing, such as molecular nitrogen and carbon dioxide, were probably also present. Strongly oxidizing species such as molecular oxygen were presumably absent.

It is also likely that there existed on the primitive earth substantial quantities of heavier hydrocarbon molecules and other reduced organic species which would have either dissolved in the primitive oceans or formed an oily scum on their surface. Heavier molecules of this sort are seen in low abundance in the sky and in greater abundance in carbonaceous meteorites. The geophysical stage of pre-biotic evolution thus provides us with a wide choice of hydrogen-rich gases and liquids to serve as the input for the chemical stage.

The chemical stage of pre-biotic evolution was explored in the classic experiments of Miller, in 1953, and in many later experiments (Miller & Orgel, 1974). You are probably familiar with the Miller experiments. Miller took a plausible reducing atmosphere composed of methane, ammonia, molecular hydrogen and water, passed electric sparks through it and collected the reaction products. He found a mixture of organic compounds containing a remarkably high fraction of amino-acids. In particular he found a 2 per cent yield of alanine. Glycine and alanine are the simplest of the twenty amino-acid building-blocks out of which all living creatures build proteins. Miller also found that when he added hydrogen sulphide to his atmosphere he obtained respectable yields of the essential sulphur-containing amino-acids methionine and cysteine. The experiment worked almost equally well with an atmosphere of molecular hydrogen, molecular nitrogen and carbon monoxide. It failed completely in an oxidizing atmosphere containing either free oxygen or sulphur dioxide. Other people have repeated the Miller experiments with many variations, using ultraviolet light or ionizing radiation as the energy-source instead of electric sparks. The results are always consistent. The input of energy into a reducing atmosphere causes the production of amino-acids in substantial quantity. The input of energy into an oxidizing atmosphere does not.

The upshot of the Miller experiments is that we can rely

21

on nature to provide an ample supply of amino-acids on the primitive earth. So long as there was a reducing atmosphere with plenty of ultraviolet light pouring in from the sun, amino-acids must have been made in large quantities. They would have descended from the atmosphere with rainfall and would have accumulated in land-locked lakes and seas where rainwater evaporates and leaves its dissolved chemicals behind. This is a familiar story, ending with a picture of a pond full of rich warm soup on the lifeless earth, the pond containing a concentrated solution of amino-acids and other organic molecules and waiting only for the breath of life to stir them into organized activity. So far as the geophysical and chemical stages are concerned, this story is probably a good approximation to the truth.

The pre-biotic synthesis of nucleotides is a more difficult problem. Efforts to synthesize nucleotides directly from their elementary components in a Miller-style experiment have not been successful. A nucleotide is a lop-sided molecule made up of three parts, an organic base plus a sugar plus a phosphate ion. The most plausible way to synthesize an organic base was demonstrated in an experiment by Oró in 1960. Oró prepared a concentrated solution of ammonium cyanide in water and simply let it stand. He found that the ammonium cyanide was converted into the organic base adenine with 0·5 per cent yield. Adenine is one of the four bases, adenine, thymine, guanine and cytosine, which are the active ingredients of DNA. Oró was also able to synthesize guanine in a similar way. The starting material, ammonium cyanide, is likely to have been present in solution in the surface waters of the earth so long as the reducing atmosphere lasted. It is not easy to imagine an ammonium cyanide solution under natural conditions becoming sufficiently concentrated to make the Oró synthesis occur with high yield. Leslie Orgel has suggested one possible way in which this might have happened. If a pond containing a dilute solution of ammonium cyanide freezes, the ice on top will be almost pure water and the solution in the unfrozen liquid below will become more concentrated as the freezing proceeds. If the temperature falls slowly and the freezing continues smoothly, the final result will be a small volume of concentrated eutectic solution of ammonium cyanide at the bottom of the pond. The temperature at which the eutectic solution finally freezes is minus 22 °Celsius. Conceivably, the concentrated solution at a temperature around minus 20 °Celsius might remain undisturbed for a long enough time to produce adenine by the Oró reaction. As Leslie Orgel has remarked, what we need to give us a natural synthesis of nucleotide bases is not a warm soup but a very cold soup. The Oró synthesis of nucleotide bases requires much more special conditions than the Miller synthesis of amino-acids. The sugar component of nucleotides can also be synthesized with reasonable efficiency, from a concentrated solution of formaldehyde. Formaldehyde is a molecule which is seen occurring naturally in molecular clouds in the sky. Like the Oró synthesis, the sugar synthesis requires special conditions to bring formaldehyde to a sufficiently high concentration. The phosphate ion, which is the third component of nucleotides, does not need to be synthesized, since it occurs naturally as a mineral constituent in rocks and sea-water.

We have thus found possible mechanisms for the natural synthesis of each of the three parts of a nucleotide. The major difficulty arises when we try to find a natural way to stick the three parts together in the right geometrical arrangement. If the linkages are made at random, only about one in one hundred molecules will be stereochemically correct. It is difficult to imagine a natural process which could separate a correctly formed nucleotide from its ninety-nine misshapen brothers. And finally, the correctly formed nucleotides are unstable in solution and tend to hydrolyse back into their components. We cannot assume that nucleotides continued to accumulate in primitive ponds for thousands of years. The rate of synthesis of nucleotides must be high to keep pace with the rate of hydrolysis. The nucleotides in our bodies are stable only because they are packaged in double helices which protect them from hydrolysis. The nucleotides on the primitive earth would have been rare birds, difficult to synthesize and easy to dissociate. Nobody has yet discovered a way to make them out of their components rapidly enough so that they would have a reasonable chance of finding each other and combining into stable helices before they hydrolysed.

The results of thirty years of intensive chemical experimentation have shown that the pre-biotic synthesis of amino-acids is easy to simulate but the pre-biotic synthesis of nucleotides is not. We cannot say that the pre-biotic synthesis of nucleotides is impossible. We know only that if it happened it happened by some process which none of our chemists has been clever enough to reproduce. This conclusion may be considered to favour the double-origin hypothesis and to argue against a single-origin hypothesis for the origin of life. A single-origin hypothesis requires both amino-acids and nucleotides to be synthesized by natural processes before life began. The double-origin hypothesis requires only amino-acids to be synthesized pre-biotically, the nucleotides being formed later as a by-product of protein metabolism. The evidence from chemical simulations does not disprove the single-origin hypothesis but it makes at least a strong presumptive case against it.

Genetics and paleontology

I have discussed at some length the evidence about the origin of life provided by experimental chemistry. I will now discuss more briefly the evidence from genetics and paleontology. The main fact that we have learned from modern genetics is that the genetic apparatus is universal. By the genetic apparatus I mean the organization of ribosomes and transfer RNA molecules which enables a cell to translate a nucleic acid gene into a protein according to the triplet code. In modern cells the gene is transcribed into a molecule of messenger RNA before being translated. The forms of messenger RNA are highly variable from species to species. The triplet code is embodied in the transfer RNA molecules and is the same in all cells apart from some minor differences. This universality of the genetic apparatus is strong evidence that all existing cells are descended from a common ancestor.

And there must have been a unique latest common ancestor, a single cell whose progeny diversified into the myriad branches of the evolutionary tree. We know from the genetic evidence that the latest common ancestor already possessed an essentially complete genetic apparatus with the same triplet code which is used in modern cells.

George Fox and Carl Woese have been the chief explorers of the earliest branchings of the evolutionary tree. They delineated the early branches by measuring quantitatively the degree of relatedness of the nucleotide sequences of ribosomal RNA molecules in widely diverse cells. The ribosomal RNA molecules, being crucial to the functioning of the genetic apparatus, are intensely conservative and change their sequences extremely slowly. Nevertheless they do change, and the divergence between sequences in two cells measures in a rough fashion the time which has elapsed since those two cells shared a common ancestor. Tracing the relationships between cells in this way, one finds that the evolutionary tree has three main branches representing a divergence of cell-types far more ancient than the later division of creatures into animals and plants. Moreover, the genetic apparatus carried by organelles such as chloroplasts and mitochondria within eucaryotic cells does not belong to the same main branch of the tree as the genetic apparatus in the nuclei of the eucaryotic cells. Terzaghi *et al.* (1984) have published a useful collection of papers summarizing the evidence of early evolution which has been derived from the study of molecules. The difference in genetic apparatus between organelles and nucleus is the strongest evidence confirming Lynn Margulis's theory that the organelles of the modern eucaryotic cell were originally independent free-living cells and only later became parasites of the eucaryotic host. According to this theory the evolutionary success of the eucaryotic cell was due to its policy of free immigration. Like the United States of America in the nineteenth century, the eucaryotic cell gave shelter to the poor and homeless and exploited their talents for its own purposes. Needless to say, both in the United States and in the eucaryotic cell, once the old immigrants are comfortably settled and their place in society is established, they do their best to shut the door to any prospective new immigrants.

Another significant fact which emerged from recent experiments is that there are some differences not only in the sequences of ribosomal RNA molecules but in the genetic code itself between mitochondria and independently living cells. The mitochondria of various species have minor variations of the code which are not seen elsewhere. This fact is additional evidence for the parasitic origin of mitochondria. Nevertheless, the fact that differences exist is less important than the fact that differences are very slight. Even in the most striking cases, the mitochondrial code is so close to the standard code that the two codes cannot be genetically unrelated. The exceptions to the universality of the code do not weaken the case for believing that the mitochondrion and its host, for however long a time they may have been separated, were originally descended from a common ancestor.

This argument for the existence of a common ancestor applies only to cells and organelles which possess a genetic translation apparatus. It does not apply to creatures such as viruses which reproduce only within cells and borrow the genetic apparatus of the cells which they invade. The genetic structures of viruses do not give us direct evidence of their antiquity. Viruses may be very ancient or they may have originated comparatively recently as plasmids, that is to say as pieces of nucleic acid detached from the chromosomes of normal cells and pursuing a more or less independent existence within the cells. A virus may be nothing more than a plasmid which has learned to survive outside its host cell by covering itself with a protein coat. Alternatively, a virus may be a highly degenerate descendant of a normal cell which has adopted a parasitic mode of life and lost all of its metabolic functions. The origin of viruses is still an open question. All that we can say for sure is that, since viruses as they now exist are totally parasitic, there must have been cells before there were viruses. There is no way in which we can imagine a virus coming first and later growing into a cell.

We know even less concerning the possible origin of a

mysterious group of organisms which have recently been given the name of 'prions'. These are the organisms that are responsible for some slow degenerative diseases of the central nervous system, scrapie in sheep, kuru in humans and possibly others. The name 'prion' was given to them by Stanley Prusiner (Prusiner, 1982), who has been working for many years on the chemical analysis of the scrapie agent. Almost nothing is known about the structure of prions, and little about their chemistry. All that is known for certain is that they multiply and cause disease within the brains of animals, and that they are unique among living creatures in giving no positive response to any of the standard chemical tests for the presence of nucleic acid. It is conjectured, but not proved, that they are creatures made of pure protein without any nucleic acid at all. How they might succeed in reproducing without nucleic acid is still a mystery. Presumably they have found a way to invade nerve-cells and subvert the nucleic acid in the nuclei of the cells into organizing the manufacture of copies of themselves. For this to be possible they must be made of the same kind of protein which exists naturally in nerve-cells. The structure and life-cycle of prions are now being actively investigated in many places, and with luck within a few years we shall understand prions as well as we understand viruses. The understanding of their structure may or may not lead to an understanding of their origin.

I am putting forward in these lectures such evidence as I can collect to support the hypothesis that life had a double origin. The double-origin hypothesis implies that the first living creatures were able to metabolize but not to replicate and that they were built mainly out of protein rather than nucleic acid. You might at this point expect me to claim that the existence of the prion is evidence confirming this hypothesis. I make no such claim. The prion would be confirming evidence only if it could be proved to be a primeval relic directly descended from the earliest creatures which lived before the development of the nucleic acid genetic apparatus. There are several strong reasons which convince me, much as I would like to discover such a relic, that the prion cannot be primeval. The prion is like a virus in having a narrowly specialized and wholly parasitic life-cycle. Even more than a virus, it must be closely linked in chemical structure to the cells which it invades. It is difficult to imagine that a primeval cell, after lurking in odd corners of the earth for thousands of millions of years, would miraculously find itself pre-adapted to the chemistry of so sophisticated an organ as the brain of a sheep. It is far more likely that the prion originated in modern times as a misplaced fragment of a brain-cell. Another argument against the antiquity of prions is the improbability of survival of creatures without a genetic apparatus in competition with creatures of modern design. If cells without a genetic apparatus did indeed exist, they cannot have had metabolic agility, mobility and responsiveness in any way comparable with modern cells. They must have been grossly inefficient, slow and blind by modern standards. As soon as the genetic apparatus was perfected, cells possessing it had an overwhelming advantage over earlier forms of life. No matter how long the evolution of pure-protein forms of life may have lasted, and no matter how great a variety of such forms may have existed, we should not expect any living relics of that epoch to have survived. The prion is an exciting and important discovery, but it is unlikely to throw any direct light upon the question of the origin of life.

The evidence of genetic relatedness collected by Carl Woese and others proves that all existing cells have a common ancestor but does not provide an absolute date for the epoch of the latest common ancestor. Genetic evidence gives us good relative dating of the different branches of the evolutionary tree but no absolute dating. For absolute dates we must turn to the evidence of paleontology. The pioneer in the discovery of fossil evidence for the absolute dating of early life was E. S. Barghoorn. The rock in which micro-fossils are best preserved is chert, the geologists' name for the fine-grained silica rock which ordinary people call flint. Chert is formed by the slow precipitation of dissolved silica from water, a process which puts minimal stress on any small creatures that may become embedded within it. Once formed, the chert is hard and chemically inert so that fossils inside it are well protected. The microfossils which Barghoorn and others have collected are little black blobs in which internal structure is barely discernible. Not all of them are definitely known to be organic in origin. I myself cannot pretend to decide whether a microscopic blob is a fossil cell or an ordinary grain of dust. I accept the verdict of the experts who say that most of the blobs are in fact fossils.

The results of a great number of observations of micro-fossils can be briefly summarized as follows. I use the word eon to mean a thousand million years. In rocks which are reliably dated with an age of about 3 eons, mainly in South Africa, we find microfossils which resemble modern bacteria in shape and size. In rocks which are dated with an age of about 2 eons, mainly in Canada, we find fossils which resemble modern procaryotic algae, including chains of cells and other multicellular structures. In rocks which are dated with an age of 1 eon, mainly in Australia, we find fossils which resemble modern eucaryotic cells, with some traces of internal structure. The fossils are too small to be analysed chemically with any accuracy, but traces of long-chain hydrocarbons have been found in the 3-eon group, while the 1-eon group contains porphyrin residues which are presumably derived from chlorophyll.

The geological dating of the various fossil groups is remarkably accurate and reliable. Unfortunately, we do not know with equal accuracy what it is that is being dated. We do not know how to identify the various fossils with particular branches of the evolutionary tree. Except for the general similarity of size and shape, there is no evidence that the cells of the 3-eon group were cousins of modern bacteria. There is no evidence that they possessed a modern genetic apparatus with ribosomes and transfer RNA. There is no evidence of the presence of nucleic acids in any of the ancient microfossils. So far as the evidence goes, the cells of the 3-eon group may have been either cells of modern type with a complete genetic apparatus, or cells of a rudimentary kind lacking nucleic acids altogether, or anything in between. Only the cells of the 1-eon group were definitely modern with eucaryotic characteristics. So far as the evidence goes, the latest common ancestor of all living cells may have lived before the 3-eon group of fossils, or between the 3-eon group and the 2-eon group or possibly even later than the 2-eon group. The dating of the latest common ancestor requires a reliable linkage of the branch-points of the evolutionary tree with the various groups of fossils. The most urgent problem for evolutionary geneticists and paleontologists is to establish the calibration of relative ages determined by genetic linkages in terms of absolute ages determined by geology. Until this problem is solved, neither the genetic evidence nor the paleontological evidence will be sufficient to determine the date of our latest common ancestor.

Theories

We have experimental and observational evidence concerning things which happened both before and after the origin of life. Before the origin of life there were geophysical and chemical processes which left behind traces that can be observed in the earth's rocks and in the sky. After the origin of life there were evolutionary processes, which can be observed in fossils and in the taxonomy of nucleic acid molecules in living species. Concerning the origin of life itself, the watershed between chemistry and biology, the transition between lifeless organic soup and organized biological metabolism, there is no direct evidence at all. The crucial transition from disorder to order left behind no observable traces. When we try to understand the nature of this transition we are forced to go beyond experimental evidence and take refuge in theory. So I will now summarize the efforts which have been made by various people over the last sixty years to understand the transition theoretically.

There are three main groups of theories about the origin of life. I call them after the names of their most famous advocates, Oparin, Eigen and Cairns-Smith. I have not done the historical research that would be needed to find out who thought of them first. The Oparin theory was described in

Oparin's book *The Origin of Life* in 1924, long before anything was known about the structure and chemical nature of genes (Oparin, 1957). Oparin supposed that the order of events in the origin of life was: cells first, enzymes second, genes third. He observed that when a suitably oily liquid is mixed with water it sometimes happens that the two liquids form a stable mixture called a coacervate, with the oily liquid dispersed into small droplets which remain suspended in the water. Coacervate droplets are easily formed by non-biological processes and they have a certain superficial resemblance to living cells. Oparin proposed that life began by the successive accumulation of more and more complicated molecular populations within the droplets of a coacervate. The physical framework of the cell came first, provided by the naturally occurring droplet. The enzymes came second, organizing the random population of molecules within the droplet into self-sustaining metabolic cycles. The genes came third, since Oparin had only a vague idea of their function and they appeared to him to belong to a higher level of biological organization than enzymes.

The Oparin picture was generally accepted by biologists for half a century. It was popular not because there was any evidence to support it but rather because it seemed to be the only alternative to biblical creationism. Then, during the last twenty years, Manfred Eigen provided another alternative by turning the Oparin theory upside-down (Eigen *et al.*, 1981). The Eigen theory reverses the order of events. It has genes first, enzymes second and cells third. It is now the most fashionable and generally accepted theory. It has become popular for two reasons. Firstly, the experiments of both Eigen and Orgel use RNA as the working material and make it plausible that the replication of RNA was the fundamental process around which the rest of biology developed. Secondly, the discovery of the double helix showed that genes are structurally simpler than enzymes. Once the mystery of the genetic code was understood, it became natural to think of the nucleic acids as primary and of the proteins as secondary structures. Eigen's theory has self-replicating RNA at the beginning, enzymes appearing soon afterwards to build with the RNA a primitive form of the modern genetic transcription apparatus, and cells appearing later to give the apparatus physical cohesion.

The third theory of the origin of life, the theory of Cairns-Smith, is based upon the idea that naturally occurring microscopic crystals of the minerals contained in common clay might have served as the original genetic material before nucleic acids were invented (Cairns-Smith, 1982). The microcrystals of clay consist of a regular silicate lattice with a regular array of ionic sites but with an irregular distribution of metals such as magnesium and aluminium occupying the ionic sites. The metal ions can be considered carriers of information like the nucleotide bases in a molecule of RNA. A microcrystal of clay is usually a flat plate with two plane surfaces exposed to the surrounding medium. Suppose that a microcrystal is contained in a droplet of water with a variety of organic molecules dissolved in the water. The metal ions embedded in the plane surfaces form irregular patterns of electrostatic potential which can adsorb particular molecules to the surfaces and catalyse chemical reactions on the surfaces in ways dependent on the precise arrangement of the ions. In this fashion the information contained in the patterns of ions might be transferred to chemical species dissolved in the water. The crystal might thus perform the same function as RNA in guiding the metabolism of amino-acids and proteins. Moreover, it is conceivable that the clay microcrystal can also replicate the information contained in its ions. When the crystal grows by accreting silicate and metal ions from the surrounding water, the newly accreted layer will tend to carry the same pattern of ionic charges as the layer below it. If the crystal is later cut along the plane separating the old from the new material, we will have a new exposed surface replicating the original pattern. The clay crystal is thus capable in principle of performing both of the essential functions of a genetic material, that is to say replicating the information that it carries and transferring the information to other molecules. I repeat that it can do these things in principle. That is to say, it can do them with some undetermined efficiency which may be very low. There is no experimental evidence to support the statement that clay can act either as a catalyst or as a replicator with enough specificity to serve as a basis for life. Cairns-Smith asserts that the chemical specificity of clay is adequate for these purposes. The experiments to prove him right or wrong have not been done.

The Cairns-Smith theory of the origin of life has clay first, enzymes second, cells third and genes fourth. The beginning of life was a natural clay crystal directing the synthesis of enzyme molecules adsorbed to its surface. Later, the clay and the enzymes learned to make cell membranes and became encapsulated in cells. The cells contained clay crystals performing in a crude fashion the functions performed in a modern cell by nucleic acids. This primeval clay-based life may have existed and evolved for many millions of years. Then one day a cell made the discovery that RNA is a better genetic material than clay. As soon as RNA was invented, the cells using RNA had an enormous advantage in metabolic precision over the cells using clay. The clay-based life was eaten or squeezed out of existence and only the RNA-based life survived.

At the present time there is no compelling reason to accept or to reject any of the three theories. Any of them, or none of them, could turn out to be right. We do not yet know how to design experiments which might decide between them. I happen to prefer the Oparin theory, not because I think it is necessarily right but because it is unfashionable. In recent years the attention of the experts has been concentrated upon the Eigen theory, and the Oparin theory has been neglected. The Oparin theory deserves a more careful analysis in the light of modern knowledge. In Chapter 3 I describe my own attempt to put the Oparin theory into a modern framework using the mathematical methods of Kimura.

There are numerous other theories which I have no time to discuss in detail. In particular, my physicist colleague Philip Anderson at Princeton has recently proposed a theory which uses the concept of a spin-glass in solid-state physics as a model of the transition to biological order (Anderson,

1983). I group Anderson's model together with Eigen's since both of them are primarily concerned with the replication of nucleic acids. I confine my discussion here to the theories of Oparin, Eigen and Cairns-Smith. Much of what I say about Eigen applies also to Anderson.

I find it illuminating to look at these theories in the light of the question that I raised in Chapter 1, whether the origin of life was a single or a double process, that is to say whether metabolism and replication originated together or separately. The Cairns-Smith theory is explicitly a double-origin theory. That to my mind is its main virtue. It has the first origin of life mainly concerned with the building of a protein metabolic apparatus, the clay particles adding to this apparatus a replicative element which may or may not be essential. The second origin of life, which Cairns-Smith calls 'Genetic takeover', is the replacement of the clay component by an efficient replicative apparatus made of nucleic acids. Cairns-Smith imagines the two origins of life to be separated by a long period of biochemical evolution, so that the nucleic acid invasion occurs in cells already highly organized with protein enzymes and lipid membranes. The Oparin theory and the Eigen theory were presented as single-origin theories. Each of them supposes the origin of life to have been a single process. Oparin places primary emphasis on metabolism and barely discusses replication. Eigen places primary emphasis on replication and imagines metabolism falling into place rapidly as soon as replication is established. I am suggesting that the Oparin and Eigen theories make more sense if they are put together and interpreted as the two halves of a double-origin theory. In this case, Oparin and Eigen may both be right. Oparin is describing the first origin of life and Eigen the second. With this interpretation, we combine the advantages of the two theories and eliminate their most serious weaknesses. Moreover, the combination of Oparin and Eigen into a double-origin theory is not very different from the theory of Cairns-Smith. Roughly speaking, Cairns-Smith equals Oparin plus Eigen plus a little bit of clay. All three theories may turn out to contain essential elements of the truth.

The error catastrophe

I shall not discuss the Cairns-Smith theory any further. I have nothing new to say about the adequacy or inadequacy of clay as a genetic material. I will talk about the Oparin and Eigen theories as if these were the only alternatives. You may regard the phrase 'the Eigen theory' as shorthand for 'any theory of the origin of life which puts replication first', and 'the Oparin theory' as shorthand for 'any theory of the origin of life which puts metabolism first'.

Before going into detail with my own version of the Oparin theory I shall describe the main features of the Eigen theory. The Eigen theory is based on two concepts which he calls quasi-species and hypercycles. The quasi-species comes first and is concerned only with replication. The hypercycle comes a short time later and introduces a metabolic system coupled to the replicative system. Eigen supposes that life began with an ample supply of nucleotide monomers capable of polymerizing into RNA as they did in his laboratory experiments. A quasi-species is a population of genetically related but not identical RNA molecules. The molecules in a quasi-species form templates for the formation of new RNA molecules which also belong to the quasi-species. Since the replication process is not perfect, the molecules will mutate and diversify from generation to generation. Nevertheless, Eigen assumes that the variations in molecular structure within a quasi-species remain bounded and settle down to a steady state. The population of molecules in a quasi-species shares a common morphology, like the population of individual organisms in a biological species. For this common morphology to persist in a steady state, Eigen assumes that there is a Darwinian process of selection favouring the replication of molecules which come close to the quasi-species norm. Eigen describes this situation quantitatively with a simple set of equations representing the balance between Darwinian selection and random errors of replication.

The hypercycle is a higher level of organization which comes into existence when several quasi-species of RNA are established and have begun to form chemical associations with friendly populations of protein enzymes. The enzymes associated with one quasi-species are supposed to assist the replication of a second quasi-species, and vice versa. The linked populations then become locked into a stable equilibrium.

Even at the quasi-species level of the Eigen theory there is a serious difficulty. The central problem for any theory of the origin of replication is the fact that a replicative apparatus has to function almost perfectly if it is to function at all. If it does not function perfectly, it will give rise to errors in replicating itself, and the errors will accumulate from generation to generation. The accumulation of errors will result in a progressive deterioration of the system until it is totally disorganized. This deterioration of the replication apparatus is called the 'error catastrophe'.

Manfred Eigen has given us a simple mathematical statement of the error catastrophe as follows. Suppose that a self-replicating system is specified by N bits of information and that each time a single bit is copied from parent to daughter the probability of error is ε. Suppose that natural selection operates to penalize errors by a selection factor S. That is to say, a system with no errors has a selective advantage S over a system with one error, and so on. Then Eigen finds the criterion for survival to be

$$N\varepsilon < \log S. \qquad (2.1)$$

If the condition (2.1) is satisfied, the selective advantage of the error-free system is great enough to maintain a population with few errors. If the condition (2.1) is not satisfied, the error catastrophe occurs and the replication cannot be sustained. The meaning of this condition (2.1) is easy to interpret in terms of information theory. The left side ($N\varepsilon$) of the inequality is the number of bits of information lost by copying errors in each generation. The right side ($\log S$) is the number of bits of information supplied by the selective action of the environment. If the information supplied is less than the information lost in each generation, a progressive degeneration is inevitable.

The condition (2.1) is very stringent. Since the selective advantage of an error-free system cannot be astronomically large, the logarithm cannot be much greater than unity. To satisfy this condition (2.1) we must have an error-rate of the order of N^{-1} at most. It is barely satisfied in modern higher organisms which have N of the order of 10^8 and ε of the order of 10^{-8}. To achieve an error-rate as low as 10^{-8}, modern organisms have evolved an extremely elaborate system of double-checking and error-correction within the replication system. Before any of this delicate apparatus existed, the error-rates must have been much higher. The condition (2.1) thus imposes severe requirements on any theory of the origin of life which, like Eigen's theory, makes the replication of RNA a central element of life from the beginning.

All the experiments which have been done with RNA replication under abiotic conditions give error-rates of the order of 10^{-2} at best. If we try to satisfy the condition (2.1) without the help of pre-existing organisms, we are limited to a replication system that can describe itself with less than one hundred bits of information. One hundred bits of information are far too few to describe any interesting protein chemistry. This does not mean that Eigen's theory is untenable. It means that Eigen's theory requires an information-processing system which is at the same time extraordinarily simple and extraordinarily free from error. We do not know how to achieve such low error-rates in the initial phases of life's evolution.

I chose to study the Oparin theory because it offers a possible way of escape from the error catastrophe. In the Oparin theory the origin of life is separated from the origin of replication. The first living cells had no system of precise replication and could therefore tolerate high error-rates. The main advantage of the Oparin theory is that it allows early evolution to proceed in spite of high error-rates. It has the first living creatures consisting of populations of molecules with a loose organization and no genetic fine-tuning. There is a high tolerance for errors because the metabolism of the population depends only on the catalytic activity of a majority of the molecules. The system can still function with a substantial minority of ineffective or uncooperative molecules. There is no requirement for unanimity. Since the statistical fluctuations in the molecular populations will be large, there is a maximum opportunity for genetic drift to act as the driving force of evolution.

To end this chapter, I briefly describe the results of a penetrating study of the Eigen theory carried out by Ursula Niesert and her colleagues at Freiburg in Germany (Niesert *et al.*, 1981). The title of their paper is 'Origin of life between Scylla and Charybdis'. They ran a large number of computer simulations of quasi-species and hypercycles behaving according to Eigen's rules. Niesert discovered that the error catastrophe is not the only catastrophe to which Eigen's molecular populations may succumb. She found three other catastrophes, each occurring with high frequency in her computer runs and each having an obvious biological interpretation. The three additional catastrophes were called selfish RNA, short-circuit and population collapse. The selfish RNA catastrophe occurs when a single RNA molecule, as a result of a sequence of mutations, learns to replicate itself faster than its competitors but forgets its function as a catalyst. The selfish RNA then becomes a parasite and quickly chokes the rest of the population to death. The short-circuit catastrophe occurs when an RNA molecule, which is supposed to be a link in the chain of hypercycle reactions, changes its sequence in such a way as to catalyse a later reaction in the chain. The chain is then short-circuited and the hypercycle contracts to a simpler hypercycle or to a single cycle. The population collapse catastrophe occurs when, as a result of a statistical fluctuation, the population of molecules in one of the essential components of a hypercycle falls to zero. The entire hypercycle then rapidly collapses.

Niesert found in her computer simulations that the probability of a selfish-RNA catastrophe or a short-circuit catastrophe increases with the size of the molecular population. Since a single aberrant molecule causes the catastrophe, the increase of probability with population-size is to be expected. In order to escape the selfish-RNA and short-circuit catastrophes for a reasonable length of time, the population of a hypercycle model must be kept small. On the other hand, the probability of the population collapse catastrophe is large for small populations and decreases as the population-size increases. Consequently, the hypercycle model must sail carefully between the Scylla of selfish-RNA and short-circuit and the Charybdis of population collapse. There is only a narrow range of population-size for which all three calamities are unlikely, and even at the optimum population-size the lifetime of the hypercycle is finite.

Niesert's results are important because they reveal perils which are likely to exist in any model of the early stages of the development of life. Early life, whether or not it is correctly described by Eigen's hypercycle model, had to sail between the perils of Scylla and Charybdis. Niesert's analysis is not merely a criticism of the Eigen theory. It is a criticism of any theory of the origin of life which assumes a cooperative organization of a large population of molecules without providing explicit safeguards against short-circuiting of metabolic pathways. The Oparin theory has not yet been tested by computer simulations in the style of Niesert. Until it has been so tested, we have no right to assume that it can deal with Niesert's three catastrophes any better than the Eigen theory can.

3   A toy model

The meaning of metabolism

This chapter describes my own attempt to understand the Oparin theory of the origin of life. The Oparin theory as Oparin proposed it made no attempt to be quantitative. I am trying to place the theory within a framework of strict mathematics so that its consequences can be calculated. The essential difficulty arises from the fact that metabolism is a vague and ill-defined concept. There is no such difficulty with the concept of replication. Replication means exactly what it says. To replicate a molecule means to copy it, either exactly or with a stated margin of error. Starting from this well-defined concept, Manfred Eigen was able to formulate his theory of the origin of life, which is in fact a theory of the origin of replication, as a system of equations which can be solved with a computer. Eigen's equations describe the evolution with time of populations of molecules subject to non-linear laws of replication. When we try to formulate in a similarly exact fashion the theory of Oparin, which is a theory of the origin of metabolism, we run immediately into the problem of defining what we mean by metabolism.

I have tackled this problem in two stages. In the first stage I have set up a mathematical description of molecular populations in the fashion in which physicists are accustomed to describe classical dynamical systems, making the dynamical equations precise but leaving the laws of interaction completely general. The general theory of molecular systems obtained in this way allows us to define what we mean by the origin of metabolism but does not allow us to predict under what conditions metabolism will occur. The second stage consists of the reduction of the general theory to a toy model by the assumption of a simple and arbitrary rule for the probability of molecular interactions. The entire intricate web of biochemical processes is replaced in the model by a single formula. The habit of constructing toy models of this sort is one to which theoretical physicists easily become addicted. When the real world is recalcitrant, we build ourselves toy models in which the equations are simple enough for us to solve. Sometimes the behaviour of the toy model provides illuminating insight into the behaviour of the real world. More often, the toy model remains what its name implies, a plaything for mathematically inclined physicists. In the present case, the toy model may have some connection with reality or it may not. Whether or not its premises are reasonable, at least its conclusions are definite. Given the premises from which it starts, it behaves as one would wish a primeval molecular population to behave, jumping with calculable probability between two states which differ by the presence or absence of metabolic organization. But before giving you the detailed description of the toy model, I go back to stage one and define the general theory of molecular evolution of which the toy model is a special case.

The general theory begins with an abstract multidimensional space of molecular populations. Each point of the space corresponds to a particular list of molecules which are supposed to be present at a particular moment in a particular population. The population is confined in a droplet as Oparin imagined it. Small molecules that are free to diffuse from the surrounding medium into the droplet and out again are not counted. The population of molecules within the droplet can change from moment to moment, either by chemical reactions within the population, or by reactions incorporating small molecules from the medium or by reactions rejecting small molecules into the medium. The theory represents all these chemical reactions by a single matrix M of probabilities. Given a population in the state A, the next chemical reaction will take it to the state B with a certain probability which is the matrix element of M between the states A and B. The population thus evolves in a stepwise and stochastic fashion over the space of possible states. The stepwise evolution can be described by a simple linear equation

$$P(k+1) = M P(k), \qquad (3.1)$$

where $P(k)$ is the probability-distribution of the population after k steps and $P(k+1)$ is the probability-distribution after $(k+1)$ steps. This equation has the formal solution

$$P(k) = M^k P(0), \qquad (3.2)$$

where $P(0)$ is the probability-distribution of populations in an arbitrary initial state. We are interested in population-distributions that persist during evolution over long periods of time. We call such distributions quasi-stationary. A quasi-stationary distribution Q satisfies an equation

$$MQ - Q = F, \qquad (3.3)$$

where F is the flow-vector describing the leakage of populations out of the distribution Q, and F is small, of the order of n^{-1} for a distribution that persists through n chemical reactions.

There will in general be a finite number of disjoint quasi-stationary distributions of various degrees of longevity. Each quasi-stationary distribution has a basin of attraction, a region of the configuration-space within which all populations are attracted to it if allowed to evolve for a long time. The various basins of attraction cover the space without overlapping. Each quasi-stationary distribution Q has a leakage-flow F which defines the small probability of a population in the basin of attraction of Q escaping over the watershed into some neighbouring basin.

Metabolism, unlike replication, is not a uniquely defined chemical process. Metabolism is a matter of degree. Some quasi-stationary distributions of molecules will metabolize more than others. Even a completely random and disorganized population of molecules will metabolize to some extent when placed in an environment providing a supply of fresh molecular components. The essential feature that we are looking for in a model of the origin of metabolism is a molecular system with two or more basins of attraction separated by high barriers. Each basin will then have a distinct quasi-stationary distribution within it. If we choose a random assortment of molecules and allow it to evolve stochastically in accordance with Eqn (3.1), it will generally end up in a particular quasi-stationary distribution which we may call the dead or disorganized state. Other quasi-stationary states that are separated from the disorganized state by high barriers are likely to be more structured, possibly with active biochemical cycles and higher rates of metabolism.

The decisive events in a theory of the origin of metabolism are the rare statistical jumps when a molecular population in one quasi-stationary state happens to undergo a succession of chemical reactions which push it up against the gradient of probability over a barrier and down into another quasi-stationary state. If the initial state is disorganized and the final state is organized, the jump may be considered to be a model for the origin of metabolism. In a complete theory of the origin of life it is likely that there would be several such jumps, each jump taking a population of molecules to a new quasi-stationary state, each quasi-stationary state having a more complex structure than the previous state. In my toy model I shall be concerned only with the first jump. In order to describe the first jump it is sufficient to have a model with two quasi-stationary states, one disordered and the other to some extent ordered or biochemically active. The barrier between the two states must be high enough to give the ordered state reasonable longevity but not so high as to make the jump from disorder to order practically impossible. We require the probability of the jump to be negligible for an individual droplet but still large enough so that jumps will occur occasionally in an assemblage of many droplets existing for a long period of time.

This is the end of stage one in my discussion of the origin of metabolism, the stage of abstract mathematical description without a specific model. The major obstacle that we face in constructing a realistic theory is the fact that we have no experimental information about possible metabolic cycles that are substantially simpler than the very complicated cycles we see in modern organisms. The primeval metabolic cycles must have been simpler than the modern ones but we do not know what they were. We do not even have plausible candidates for the rudimentary enzymes which must have been the ultimate ancestors of modern enzymes. Lacking such plausible candidates, we cannot begin to calculate the probability that a population of molecules would make the jump into a self-sustaining metabolic cycle. But my intention is to construct a model that is specific enough for me to describe the jump in detail and calculate the probabilities. I therefore leave aside the definition of the primeval metabolic cycles as a job for the future. It is a job for chemists rather than for physicists. I now proceed to stage two of my theory. I give a description of a model of the origin of metabolism without stopping to define what metabolism means. You can judge for yourselves whether the assumptions of the model are plausible. The chief virtue of the model is that its consequences are calculable. It makes quantitative and not altogether trivial statements about the molecules which might be capable of making the jump from disorder to metabolic activity.

Details of the model

The model that I construct below is called a Toy Model of the Oparin Theory (Dyson, 1982). The model is not intended to be realistic. It leaves out all the complicated details of real organic chemistry. Its purpose is to provide an idealized picture of molecular evolution that resembles in some qualitative fashion the Oparin picture of the origin of life. After I have described the toy model and deduced its consequences, I will return to the question of whether the behaviour of the model has any relevance to the evolution of life in the real world. The model is an empty mathematical frame into which we may later try to fit more realistic descriptions of pre-biotic evolution. My analysis of the model is an elementary exercise in population biology, using equations borrowed from Fisher and Kimura. The equations

44

are the same whether we are talking about a population of molecules in a droplet or a population of birds on an island (Kimura, 1970).

To define the model, I make a list of ten assumptions. The list begins with general statements but by the end the model is uniquely defined. This makes it easy to generalize the model by modifying only the more specific assumptions.

Assumption 1 (*Oparin Theory*). Cells came first, enzymes second, genes much later.

Assumption 2. A cell is an inert droplet containing a population of polymer molecules that are confined to the cell. The polymers are composed of monomer units which we may imagine to be similar to the amino-acids that make modern proteins. The polymers in the cell contain a fixed number, N, of monomers. In addition there is an external supply of free monomers that can diffuse in and out of the cell, and there is an external supply of energy which causes chemical reactions between polymers and monomers.

Assumption 3. Cells do not die and do not interact with one another. There is no Darwinian selection. Evolution of the population of molecules within a cell proceeds by random drift.

Assumption 4. Changes of population occur by discrete steps, each step consisting of a single substitution mutation. A mutation is a replacement of one monomer by another at one of the sites in a polymer. This assumption is unnecessarily restrictive and is imposed only for the sake of simplicity. At the cost of some complication of the mathematics, we could include a more realistic variety of chemical processes such as splitting and splicing of polymer chains or addition and subtraction of monomers.

Assumption 5. At every step, each of the N sites in the polymer population mutates with equal probability ($1/N$). This assumption is also unrealistic and is made to keep the calculations simple.

Assumption 6. In a given population of polymers, the bound monomers can be divided into two classes, active and inactive. This assumption appears to be uncontroversial but it actually contains the essential simplification which makes

45

the model mathematically tractable. It means that we are replacing the enormous multidimensional space of molecular configurations by a single variable, taking only two values, one for 'active' and zero for 'inactive'.

Assumption 7. The active monomers are in active sites where they contribute to the ability of a polymer to act as an enzyme. To act as an enzyme means to catalyse the mutation of other polymers in a selective manner so that the correct species of monomer is chosen preferentially to move into an active site.

Assumption 7 is the place where the notion of metabolism sneakily enters the model without being defined. The definition of the word 'active' in Assumption 7 is circular. An active monomer is one which helps other monomers to be active. So far as the model is concerned, metabolism means the cyclic shuffling around of monomer units, and the population as a whole is metabolically active if the cyclic shuffling maintains the active monomers at a self-sustaining high level.

Assumption 8. In a cell with a fraction x of monomers active, the probability that the monomer inserted by a fresh mutation will be active is $\phi(x)$. The function $\phi(x)$ represents the efficiency of the existing population of catalysts in promoting the formation of a new catalyst. The assumption that $\phi(x)$ depends on x means that the activity of catalysts is to some extent inherited from the parent population by the newly mutated daughter. The form of $\phi(x)$ expresses the law of inheritance from parent to daughter. The numerical value of $\phi(x)$ will be determined by the details of the chemistry of the catalysts.

Assumption 8 is a drastic approximation. It replaces the average of the efficiencies of a population of catalysts by the efficiency of an average catalyst. I call it the 'mean field approximation' since it is analogous to the approximation made in the Curie-Weiss mean-field model of a ferromagnet. In physics, we know that the mean-field approximation gives a good qualitative account of the behaviour of a ferromagnet. In population biology, similar approximations have been made by Kimura. The effect of the mean-field approximation

46

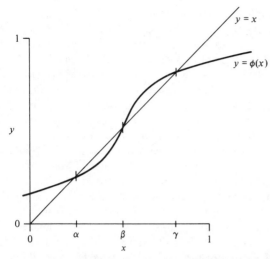

Fig. 3. The S-shaped curve $y = \phi(x)$. The curve crosses the line $y = x$ at three points which represent possible equilibrium populations. The upper and lower equilibria are stable, the middle equilibrium unstable.

is to reduce the multidimensional random walk of molecular populations to a one-dimensional random walk of the single parameter x. Both in physics and in population biology, the mean-field approximation may be described as pessimistic. It underestimates the effectiveness of local groupings of molecules in forming an ordered state. The mean-field approximation generally predicts a lower degree of order than is found in an exact theory.

Assumption 9 (*Fig. 3*). The curve $y = \phi(x)$ is S-shaped, crossing the line $y = x$ at three points, $x = \alpha, \beta, \gamma$, between zero and one. This assumption is again borrowed from the Curie-Weiss model of a ferromagnet. It means that the population of molecules has three possible equilibrium states. An equilibrium state occurs whenever $\phi(x) = x$ since the law of inheritance then gives a daughter population with the same average activity x as the parent population. The equilibrium is stable if the slope of the curve $y = \phi(x)$ is less than unity, unstable if the slope is greater than unity. Consider, for example, the lowest equilibrium state $x = \alpha$.

I call it the disordered state because it has the smallest average activity. Since the slope at α is less than unity, the equilibrium is stable. If a parent population has its average activity *x* a little above α, the daughter population will tend to slide back down toward α. If the parent population has *x* a little below α, the daughter population will tend to slide up toward α. The same thing happens at the upper equilibrium state *x* = γ. The upper state is also stable since the slope at γ is less than unity. I call it the ordered state because it has the largest catalytic activity. A population with activity *x* close to γ will move closer to γ as it evolves. But the middle equilibrium point *x* = β is unstable since the slope at β is greater than unity. If a population has *x* slightly larger than β, it will evolve away from β toward the ordered state *x* = γ, and if it has *x* slightly smaller than β it will slide away from β down to the disordered state *x* = α. The equilibrium at *x* = β is an unstable saddle-point.

Going back to the abstract mathematical language of stage one of my theory, we may say that the lower and upper equilibrium states of the model are two quasi-stationary states occupying two separate basins of attraction. The barrier separating the two basins is a mountain-pass with the unstable saddle-point equilibrium state at the top.

We have here a situation analogous to the distinction between life and death in biological systems. I call the ordered state of a cell 'alive' since it has most of the molecules working together in a collaborative fashion to maintain the catalytic cycles which keep them active. I call the disordered state 'dead' since it has the molecules uncoordinated and mostly inactive. A population, either in the dead or in the alive state, will generally stay there for a long time, making only small random fluctuations around the stable equilibrium. However, the population of molecules in a cell is finite and there is always the possibility of a large statistical fluctuation which takes the whole population together over the saddle-point from one stable equilibrium to the other. When a 'live' cell makes the big statistical jump over the saddle-point to the lower state, we call the jump 'death'. When a 'dead' cell makes the jump up over the saddle-point to the upper state, we call the jump 'origin of life'. Once the function $\phi(x)$ and the size N of the population in the cell are given, the probabilities of 'death' and of the 'origin of life' can easily be calculated. We have only to solve a linear difference equation with the appropriate boundary conditions to represent an ensemble of populations of molecules diffusing over the saddle-point from one side or the other.

The probability of the transition over the saddle-point is given precisely by the flow term F in Eqn (3.3) of the general mathematical description that I gave earlier. It is perhaps misleading to use the word 'jump' to describe the transition. The word 'diffusion' describes it better. The population of molecules climbs up the barrier by a succession of numerous small steps rather than by one big step. The population will probably take nine steps backward for every ten steps forward as it inches its way uphill. Fortunately, the net flow F can easily be calculated without following the details of the random walks by which the flow is achieved.

Assumption 10. Here we make a definite choice for the function $\phi(x)$, basing the choice on a simple thermodynamic argument. It will turn out happily that the function $\phi(x)$ derived from thermodynamics has the desired S-shaped form to produce the three equilibrium states required by Assumption 9.

We assume that every catalyst in the cell works by producing a difference between the activation energies required for placing an active or inactive monomer into a mutating molecule. If the catalyst molecule is perfect, with all its monomer units active, then the difference in activation energies will be a certain quantity U which we assume to be the same for all perfect catalysts. If a catalyst is imperfect, in a cell with a fraction x of all monomer units active, we assume that it produces a difference xU in the activation energies for correct and incorrect mutations. We are here again making a mean-field approximation, assuming that the average effect of a collection of catalysts with various degrees of imperfection is equal to the effect of a single catalyst with its discrimination reduced by the average

49

activity x of the whole population. This is another approximation that could be avoided in a more exact calculation.

We assume that the monomers belong to $(1+a)$ equally abundant species. This means that there is one right choice and a wrong choices for the monomer to be inserted in each mutation. The effect of the catalysts is to reduce the activation energy for the right choice by xU below the activation energy for a wrong choice. Thus the probability of a right choice is increased over the probability of each wrong choice by the factor

$$b^x, \qquad (3.4)$$

where

$$b = \exp(U/kT) \qquad (3.5)$$

is the discrimination factor of a perfect catalyst at absolute temperature T and k is Boltzmann's constant. We have a wrong choices with statistical weight unity compared to one right choice with statistical weight b^x. The function $\phi(x)$ is the probability of a right choice at each mutation, and therefore

$$\phi(x) = (1 + ab^{-x})^{-1}, \qquad (3.6)$$

the same S-shaped function that appears in the mean-field model of a simple ferromagnet.

The formula (3.6) for $\phi(x)$ completes the definition of the model. The model is uniquely defined once the three parameters N, a and b are chosen. The three parameters summarize in a simple fashion the chemical raw material with which the model is working. N defines the size of the molecular population, a defines the chemical diversity of the monomer units and b is the quality-factor defining the degree of discrimination of the catalysts.

We now have a definite three-parameter model to work with. We still have to calculate its consequences and to examine whether it shows interesting behaviour for any values of N, a and b that are consistent with the facts of organic chemistry. 'Interesting behaviour' here means the occurrence with reasonable probability of a jump from the disordered state to the ordered state. We shall find that interesting behaviour occurs for values of a and b lying in a narrow range. This narrow range is determined only by the mathematical properties of the exponential function and is independent of all physical or chemical constants. The model therefore makes a definite statement about the stuff out of which the first living cells were made. If the model has anything to do with reality, then the primeval cells were composed of molecules having values of a and b within the calculated range.

It turns out that the preferred ranges of values of the three parameters are:

$$a, \text{ from 8 to 10;} \tag{3.7}$$

$$b, \text{ from 60 to 100;} \tag{3.8}$$

$$N, \text{ from 2000 to 20000.} \tag{3.9}$$

The ranges also happen to be reasonable from the point of view of chemistry. (3.7) says that the number of species of monomer should be in the range from nine to eleven. In modern proteins we have twenty species of amino-acids. It is reasonable to imagine that about ten of them would provide enough diversity of protein function to get life started. On the other hand, the model definitely fails to work with $a = 3$, which would be the required value of a if life had begun with four species of nucleotides polymerizing to make RNA. Nucleotides alone do not provide enough chemical diversity to allow a transition from disorder to order in this model. The quantitative predictions of the model are thus consistent with the Oparin theory from which we started. The model decisively prefers protein to nucleic acid as the stuff from which life arose.

The range (3.8) from sixty to one hundred is also reasonable for the discrimination factor of primitive enzymes. A modern polymerase enzyme typically has a discrimination factor of 5000 or 10000. The modern enzyme is a highly specialized structure perfected by three thousand million years of fine-tuning. It is not to be expected that the original enzymes would have come close to modern standards of performance. On the other hand, simple inorganic catalysts frequently achieve discrimination factors of fifty. It is plausible that a simple peptide catalyst with an active site containing four or five amino-acids would have a discrimination factor in the range preferred by the model, from sixty to one hundred.

The size (3.9) of the population in the primitive cell is also plausible. A population of several thousand monomers linked into a few hundred polymers would give a sufficient variety of structures to allow interesting catalytic cycles to exist. A value of N of the order of 10000 is large enough to display the chemical complexity characteristic of life and still small enough to allow the statistical jump from disorder to order to occur on rare occasions with probabilities that are not impossibly small.

The basic reason for the success of the model is its ability to tolerate high error-rates. The model overcomes the error catastrophe by abandoning exact replication. It neither needs nor achieves precise control of its molecular structures. It is this lack of precision which allows a population of 10000 monomers to jump into an ordered state without invoking a miracle. In a model of the origin of life that assumes exact replication from the beginning, with a low tolerance of errors, a jump of a population of N monomers from disorder to order will occur with probability of the order of $(1+a)^{-N}$. If we exclude miracles, a replicating system can arise spontaneously only with N of the order of one hundred or less. In contrast, our non-replicating model can make the transition to order with a population that is a hundred times larger. The error-rate in the ordered state of our model is typically between 20 and 30 per cent when the parameters a and b are in the ranges (3.7), (3.8). An error-rate of 25 per cent means that three out of four of the monomers in each polymer are correctly placed. A catalyst with five monomers in its active site has one chance out of four of being completely functional. Such a level of performance is tolerable for a non-replicating system but would be totally unacceptable in a replicating system. The ability to function with a 25 per cent error-rate is the decisive factor which makes the ordered state in our model statistically accessible, with populations large enough to be biologically interesting.

Consequences of the model

The equations describing a stationary state of the model are very simple. In a quasi-equilibrium state there will be a certain probability P_j for finding j active monomers in the population. The flow-rate F in a steady state will be independent of j. The equation (3.3) defining a quasi-equilibrium state then becomes

$$f[(j+1)/N]P_{j+1} - g(j/N)P_j = F, \qquad (3.10)$$

with the functions f and g given by

$$f(x) = x[1 - \phi(x)], \; g(x) = (1-x)\phi(x). \qquad (3.11)$$

It is easy to solve Eqn (3.10) with the additional constraint that the sum of probabilities P_j must be equal to one. In this way we find an exact formula for F which is, with some unimportant approximations, equivalent to the simple formula

$$T = \tau \exp(\Delta N), \qquad (3.12)$$

for the average time T required for a cell to make the transition from disorder to order. Here τ is the average time-interval between mutations at each site, N is the total number of monomers and Δ is a number that we can calculate as it depends only on the parameters a and b. If Δ were of the order of unity, then the exponential in Eqn (3.12) would be impossibly large for N greater than one hundred. We would then be in the situation characteristic of error-intolerant systems, for which the transition to order is astronomically improbable for large N. However, when the parameters a and b are in the ranges (3.7), (3.8), which correspond to models with high error-tolerance, it turns out that Δ is not of the order of unity but lies in the range from 0·001 to 0·015. This is the feature of the model that makes transition to order possible with populations as large as 20000. Although Eqn (3.12) is still an exponentially increasing function of N, it increases much more slowly than one would naively expect.

According to Eqn (3.12) there is a critical population-size N_c such that populations N of the order of N_c or smaller will

53

make the disorder-to-order transition with reasonable probability, whereas populations much greater than N_c will not. I choose to define N_c by

$$N_c = 30/\Delta, \tag{3.13}$$

so that the exponential factor in Eqn (3.12) is

$$e^{30} \sim 10^{13} \text{ for } N = N_c. \tag{3.14}$$

The coefficient thirty in Eqn (3.13) is chosen arbitrarily. We do not know how many droplets might have existed in environments suitable for the origin of life, nor how long such environments lasted nor how frequently their molecular constituents mutated. The choice in Eqn (3.13) means that we could expect one transition to the ordered state to occur in a thousand mutation-times among a collection of 10^{10} droplets each containing N_c monomers. It is not absurd to imagine that 10^{10} droplets may have existed for a suitably long time in an appropriate environment. On the other hand, if we were to consider droplets with molecular populations three times larger, that is to say with $N = 3N_c$, then the exponential factor in Eqn (3.12) would be 10^{39}, and it is inconceivable that enough droplets could have existed to give a reasonable probability of a transition. The critical population N_c thus defines the upper limit of N for which transition can occur, with a margin of uncertainty which is less than a factor of three. The critical population-sizes given by Eqn (3.13) range from 2000 to 20000 when the parameters a and b lie in the ranges 8–10 and 60–100, respectively.

The properties of our model can be conveniently represented in a two-dimensional diagram (Fig. 4) with the parameter a horizontal and the parameter b vertical. Each point on the diagram corresponds to a particular choice of a and b. Models which satisfy the triple-crossing condition (Assumption 9) and possess disordered and ordered states occupy the central region of the diagram, extending up and to the right from the cusp. The cusp at

$$a = e^2 = 7\cdot4, \quad b = e^4 = 54\cdot6, \tag{3.15}$$

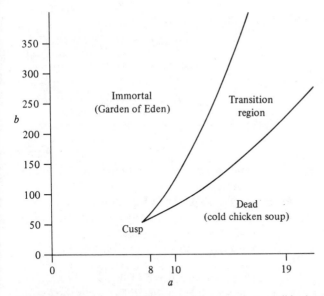

Fig. 4. Phase diagram in which each point represents a possible chemical composition of a molecular population. Plotted horizontally is a, where $(1 + a)$ is the number of species of monomers. Plotted vertically is b, the quality factor of polymer catalysts. The transition region represents populations which possess both ordered and disordered equilibrium states. The 'Dead' region has no ordered state, the 'Immortal' region no disordered state.

marks the lower bound of the values of a and b for which a disorder–order transition can occur. The critical population-size N_c is large near the cusp and decreases rapidly as a and b increase. The biologically interesting models are to be found in the part of the central region close to the cusp. These are the models that have high error-rates and can make the disorder–order transition with large populations.

To illustrate the behaviour of the model in the interesting region near the cusp, I pick out one particular case which has the advantage of being easy to calculate exactly. This is the case

$$a = 8, b = 64. \tag{3.16}$$

55

This has the three equilibrium states

$$\alpha = \tfrac{1}{3},\ \beta = \tfrac{1}{2},\ \gamma = \tfrac{2}{3}. \tag{3.17}$$

The error-rate in the ordered state is exactly one-third. The value of Δ for this model turns out to be

$$\Delta = \log 3 - (19/12) \log 2 = 0{\cdot}001129, \tag{3.18}$$

which gives a satisfactorily large critical population-size

$$N_c = 26566. \tag{3.19}$$

When I described the model at the 'From matter to life' meeting in Cambridge in 1981, my friend Christopher Longuet-Higgins, who happens to be a musician as well as a chemist, pointed out that the quantity Δ appearing in Eqn (3.18) is well known to musicians as the fractional difference in pitch between a perfect fifth and an equi-tempered fifth. On a logarithmic scale of pitch, a perfect fifth is $(\log 3 - \log 2)$ and an equi-tempered fifth is seven semitones or $(7/12) \log 2$. The smallness of the difference is the reason why the equi-tempered scale works as well as it does. The smallness of Δ is also the reason why this model of the origin of life works as well as it does. Old Pythagoras would be pleased if he could see this example, justifying his doctrine of a universal harmony that embraces number, music and science. After this digression into Pythagorean mysticism I return to the general properties of the model shown in Fig. 4.

The region below and to the right of the central strip represents models which have only a disordered state and no ordered state. These models have a too large (too much chemical diversity) and b too small (too weak a catalytic activity) to produce an ordered state. Droplets in this region are dead and cannot come to life. I call the region 'Cold chicken soup' because this phrase has been used to describe the composition of the Earth's ocean in pre-biotic times. The region above and to the left of the central strip represents models that have only an ordered state and no disordered state. These models have a too small (too little chemical diversity) and b too large (too strong a catalytic activity) to produce a disordered state. Droplets in this region are frozen into the ordered state and cannot die. I call the region 'Garden of Eden' because this phrase has been used to describe an alternative theory of the origin of life. It is possible to imagine cells evolving by random accretion of molecular components so that they drift into the central transition region either from the cold chicken soup or from the Garden of Eden. Once they reach the central region, they are capable of both life and death, and the evolution of biological complexity can begin.

As molecular populations evolve over long periods of time it is likely that both a and b will increase. An increase in a will result from incorporation of new varieties of amino-acid into the metabolic cycle. An increase in b will result from the build-up of more complicated polymer molecules able to catalyse reactions with greater specificity. Molecular populations with larger a and b will metabolize more efficiently and are likely to prevail in the Darwinian struggle for existence. Fortunately, our model allows evolution to proceed gradually upward and to the right while remaining within the critical strip of Fig. 4. It is only necessary for a and b to keep roughly in step so that b is approximately equal to the square of a. The curve $b = a^2$ remains within the critical strip all the way from the cusp to infinity. Of course, as evolution proceeds further and further up the strip, the simple assumptions of the model will represent more and more inadequately the behaviour of real organisms.

One striking feature of this model which is absent in modern organisms is the symmetry between life and death. In the model, the curve

$$y = \phi(x) = (1 + ab^{-x})^{-1} \qquad (3.20)$$

is invariant under the transformation

$$x \to 1 - x, \, y \to 1 - y, \, a \to (b/a). \qquad (3.21)$$

In particular, the model with $b = a^2$ has complete symmetry about the unstable saddle-point at $x = y = 1/2$. The ordered state and the disordered state are mirror-images of each other.

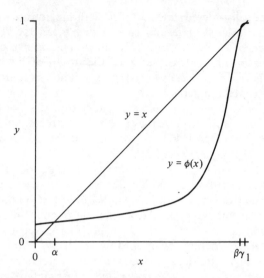

Fig. 5. The S-shaped curve of Fig. 3 as it appears in a modern cell, with the upper and middle equilibrium states pushed over far to the right so that the curve is no longer symmetrical.

The probability of a transition from disorder to order is exactly equal to the probability of a transition from order to disorder. In the symmetrical model with $b = a^2$, death and resurrection occur with equal frequency. The origin of life is as commonplace an event as death.

How did it happen that, as life evolved, death continued to be commonplace while resurrection became rare? What happened was that the catalytic processes in the cell became increasingly fine-tuned and increasingly intolerant of error. The curve $y = \phi(x)$ remained S-shaped but became more and more asymmetrical as time went on. The shape of the curve in a modern cell is shown in Fig. 5. This should be contrasted with the symmetrical curve shown in Fig. 3 for our hypothetical primitive cell. In the primitive cell the three equilibrium states might have been

$$\alpha = 0 \cdot 2, \beta = 0 \cdot 5, \gamma = 0 \cdot 8, \qquad (3.22)$$

with an error-rate of 20 per cent in the ordered state. In the modern cell the curve is pushed over far to the right and for the equilibrium state the values are typically

$$\alpha = 0.05, \beta = 0.999, \gamma = 0.9999. \qquad (3.23)$$

This value of the ordered state γ means that the error-rate in the metabolic apparatus of a modern cell is about 10^{-4}. The position of the saddle-point β means that an environmental insult such as a dose of X-rays which increases the error-rate to 10^{-3} will disrupt the fine-tuned apparatus and cause the cell to die. Death is easy and resurrection is difficult because the saddle-point has moved so close to the ordered state and so far from the disordered state. For life to originate spontaneously it was essential to have an ordered state with a high error-rate, but when life was once established the whole course of evolution was toward more specialized structures with lower tolerance of errors.

I have said enough, or perhaps too much, about the properties and the consequences of my model. You may have noticed that in talking about the model I have fallen into a trap. I have fallen in love with my model. I begin to talk about it as if it were historic truth. It is of course nothing of the kind. It is not a description of events as they really happened. It is only a toy model, a simple abstract picture which will rapidly be superseded by better models incorporating some of the chemical details that I have ignored.

4 Open questions

Why is life so complicated?

It is now time to sum up what we may have learned from the first three chapters. Chapter 1 describes the historical development of ideas leading up to the question that I consider to be fundamental to all investigation of the origin of life: Is the origin of life the same thing as the origin of replication? I give some reasons why I am inclined to answer no to this question, to give a tentative preference to the hypothesis that metabolism and replication had separate origins. Chapter 2 gives a sketchy account of some of the classic experiments and some of the classic theories concerning the origin of life. I observe that the experiments since the time of Max Delbrück have been spectacularly successful in elucidating the structure and function of the apparatus of replication and much less successful in giving us a deep understanding of metabolism. The one-sided success of the experiments has resulted in a corresponding bias of theories. The most popular theories of the origin of life are the theories of Manfred Eigen, which concentrate almost exclusively on replication as the phenomenon to be explained. Chapter 3 describes my own attempt to build a model of the origin of life with a bias opposite to Eigen's, assuming as a working hypothesis that primitive life consisted of purely metabolic machinery without replication.

This last chapter is concerned with the open questions raised by the model and with more general questions concerning possible experimental approaches to the origin of metabolism. But all these questions are subsidiary to another question: Why is life so complicated? This is perhaps not a well-posed scientific question. It might be interpreted as merely the lament of an elderly scientist recalling the lost simplicity of youth. Or it might be interpreted as an ineffectual protest against the intractability of the human condition in the modern world. But I mean the question to refer specifically to cellular structure. The essential characteristic of living cells is homeostasis, the ability to maintain a steady and more-or-less constant chemical balance in a changing environment. Homeostasis is the machinery of chemical controls and feed-back cycles which make sure that each molecular species in a cell is produced in the right proportion, not too much and not too little. Without homeostasis, there can be no ordered metabolism and no quasi-stationary equilibrium deserving the name of life. The question 'Why is life so complicated?' means in this context 'Given that a population of molecules is able to maintain itself in homeostatic equilibrium at a steady level of metabolism, how many different molecular species must the population contain?'

The biological evidence gives a rather definite answer to the question how many kinds of molecule are needed to make a homeostatic system, at least so long as we are talking about homeostatic systems of the modern type. There is a large number of different varieties of bacteria and most of them contain a few thousand molecular species if one judges the number by the few million base-pairs in their DNA. It seems that under modern conditions homeostatic systems work efficiently with a few thousand components and work less efficiently with fewer. If a bacterium could dispense with half of its molecular components and still metabolize efficiently, there would be a great selective advantage in doing so. From the fact that bacteria have generally refused to shrink below a certain level of complexity we may deduce that this level is in some sense an irreducible minimum.

If modern cells require a few thousand types of molecule for stable homeostasis, what does this tell us about primitive cells? Strictly speaking, it tells us nothing. Without the modern apparatus of genes and repressors the ancient mechanisms of homeostasis must have been very different. The ancient mechanisms might have been either simpler or more complicated. Still it is a reasonable hypothesis that the ancient mechanisms were simpler. There remains the question How simple could they have been? This question must be answered before we can build credible theories of the origin of life. It can be answered only by experiment.

In the toy model that I discuss in Chapter 3, I deduced from the arithmetic of the model that the population of a cell making the transition from disorder to order should have been between 2000 and 20000 monomers combined into a few hundred species of polymers. I claim that this number, a few hundred, was reasonable for the number of species of polymer molecule required for a primitive homeostatic system. That claim is, of course, based on nothing but guesswork. We know that a few thousand species of molecule are sufficient for a modern cell. It seems unlikely that anything resembling biochemical homeostasis could be maintained with a few tens of species. And so we guess quite arbitrarily, guided only by our familiarity with the decimal system of counting, that a few hundred kinds of molecule is the right number for the origin of homeostasis. Whether a few hundred molecular species are either necessary or sufficient for homeostasis we do not know.

It is of interest in this connection to see how an experimental approach was successfully applied to answer the corresponding question concerning the origin of replication. What is the smallest molecular population that is able to constitute a self-replicating system? This question was answered by two classic experiments, one done by Sol Spiegelman (Spiegelman, 1967), the other by Manfred Eigen and his colleagues (Eigen *et al.*, 1981). I describe Eigen's experiment in Chapter 2. Spiegelman's experiment began with a living Q_β virus, a creature able to survive and ensure its own replication in nature by means of the information coded in a single RNA molecule composed of 4500 nucleotides. The virus is normally replicated inside a host cell using a replicase enzyme which the viral RNA causes the host's ribosomes to manufacture. The viral RNA also causes the host to manufacture a coat protein and various other components that are required for the complete viral life-cycle. Now, Spiegelman proceeded to debauch the virus by stripping off its protein coat and providing it with replicase enzyme in a test-tube so that it could replicate without going to the trouble of invading a cell and completing its normal parasitic life-cycle. The test-tube also contained an ample provision of free nucleotide monomers, with a continuous-flow arrangement to keep the virus from exhausting the supply. The results were spectacular. The viral RNA continued for a while to be replicated accurately with the help of the replicase enzyme. But soon a mutant RNA appeared, having lost some of the genes that were no longer required for its survival. The mutant, having fewer than 4500 nucleotides, was replicated more quickly than the original virus and soon displaced it in the Darwinian struggle for existence. Then another still shorter mutant appeared to displace the first, and so it went on. The virus no longer needed to carry the genes for replicase and coat protein in order to survive. On the contrary, it could only survive by getting rid of all superfluous baggage. The requirement for survival was to be as simple and as small as possible. The virus finally degenerated into a little piece of RNA with only 220 nucleotides containing the recognition site for the replicase enzyme and not much else. The final state of the virus was called by geneticists the 'Spiegelman monster'. It provides a good object-lesson demonstrating what happens to you when life is made too easy. The little monsters would continue for ever to replicate at high speed in the artificial environment of Spiegelman's test-tube but could never hope to survive anywhere else.

The experiment of Manfred Eigen was the opposite of Spiegelman's experiment. Both experiments used a test-tube containing replicase enzyme and free nucleotides. Spiegelman put into this soup a living virus. Eigen put in nothing. Spiegelman was studying the evolution of replication from the top down, Eigen from the bottom up. Eigen's experiment produced a self-generated population of RNA molecules replicating with the help of the replicase enzyme, just like the Spiegelman monsters. The Eigen replicator and the Spiegelman monster were not identical but they were first cousins. Eigen's replicators, after they had evolved to a steady state, contained about 120 nucleotides each, compared with the 220 in a Spiegelman monster. The difference between 120 and 220 nucleotides is a small gap between a molecule which grew from nothing and a molecule which was once alive.

The experiments of Spiegelman and Eigen together give a clear answer to the question What is the minimum population-size required for a replicating system. The answer is, a single RNA molecule with one or two hundred nucleotides. This answer shows in a nutshell how simple the phenomenon of replication is compared with the phenomenon of homeostasis. I am conjecturing that the minimum population-size required for homeostasis would be about a hundred times larger, namely, a few hundred molecules containing ten or twenty thousand monomer units. And more importantly, I am suggesting that the most promising road to an understanding of the origin of life would be to do experiments like the Spiegelman and Eigen experiments, but this time concerned with homeostasis rather than with replication.

How could such experiments be done? I am acutely aware that it is much easier to suggest experiments than to do them. What is required first of all is to find the working materials which make experiments possible, the equivalent for a homeostatic system of Spiegelman's Q_β virus and Eigen's nucleotide soup. The objective should be once again to work from both ends, from the top down and from the bottom up, and to find out where in the middle the two ends meet. From the top, we need to find a suitable creature, an enucleated cell which has lost its replicative apparatus but still preserves the functions of metabolism and homeostasis, and we need to keep it alive artificially while stripping it gradually of inessential molecular components. We may hope in this way, with many trials and errors, to find out roughly the irreducible minimum degree of complication of a homeostatic apparatus. From the bottom, we need to experiment with synthetic populations of molecules confined in droplets in the style of Oparin, adding various combinations of catalysts and metabolites until a lasting homeostatic equilibrium is achieved. If we are lucky, we may find that the experiments from the top and those from the bottom show some degree of convergence. In so far as they converge, they will indicate a possible pathway which life might have followed in its original progress from chaos to homeostasis.

These suggestions for future experiments probably sound naive and simple-minded to experimenters whose daily lives are spent in constant battle against the recalcitrance of real cells and real chemicals. I do not know when experiments along the lines that I have suggested will become feasible. I suggest them with diffidence, being myself incapable of doing an experiment even in my own field of physics. Nevertheless, I make these suggestions with serious intent. If I did not believe that such experiments are potentially important, I would not have ventured to talk about the origin of life in the first place. If a theoretical physicist has anything of value to say about the fundamental problems of biology, it can only be through making suggestions for new types of experiment. Forty years ago, Erwin Schrödinger suggested to biologists that they should investigate experimentally the molecular structure of the gene. That suggestion turned out to be timely. I am now suggesting that biologists investigate experimentally the population structure of homeostatic systems of molecules. If I am lucky, this suggestion may also turn out to be timely.

Before leaving the subject of future experiments, I would like to add some remarks about computer simulations. In population biology the computer is a source of experimental data at least as important as field observation. Computer simulations of population dynamics are indispensable for the planning of field observations and for the interpretation of results. Computer simulations are not only quicker than field observations but also cheaper. Every serious programme of research in population biology includes computer simulations as a matter of course. Since I am advocating a programme of research in the population biology of molecules, computer simulations will be essential here too.

Ursula Niesert's computer simulations of the Eigen hypercycle model of the origin of life (Niesert *et al.*, 1981) exposed several serious weaknesses of that model. As Niesert observed as a result of her simulations, the failures of the hypercycle model are mainly due to the fact that a single RNA molecule is supposed to be performing simultaneously three separate functions. The three functions are: replicating itself with the help of a replicator molecule to which it is specifically adapted, carrying a message to promote the synthesis of another molecule and acting as agent for the specific transfer of amino-acids. The computer models show that RNA molecules have a natural tendency to specialize. They prefer to perform a single function well rather than to perform three functions badly. This conclusion is not surprising. In the natural ecology of species, it is a general rule that most species survive by becoming specialists. Niesert's simulation showed that the same rule applies in the ecology of molecules in the hypercycle model. Her criticism of the model enables us to understand it better and perhaps to improve it. In the same way, computer simulations of models of the origin of homeostasis will show us what is wrong with the models and will help us to replace them with better ones. Like the hypercycle model, models of homeostasis are likely to be vulnerable to the three dangers which Niesert described in her paper 'Origin of life between Scylla and Charybdis'. I describe the three dangers in Chapter 2. As soon as realistic models of homeostatic populations become available, computer simulations will probably reveal a variety of additional catastrophes to which they are liable. Only when we have explored all the possible modes of breakdown of homeostasis will we have a right to say that we understand what homeostasis means. Computer simulations will be essential to the growth of such understanding. In our search for an answer to the question Why is life so complicated?, biological and chemical experiments and computer simulations must always go hand in hand.

Other questions suggested by the toy model

I now return to the toy model of Chapter 3 and examine some other questions which it raises. The questions are not specific to this particular model. They will arise for any model of the origin of life in which we have molecular populations achieving metabolism and homeostasis before they achieve replication. The questions refer not to the model itself but to the implications of the model for the subsequent course of biological evolution. I comment briefly on each question in turn. After another twenty years of progress in biological research we may perhaps know whether my tentative answers are correct.

Were the first living creatures composed of proteins or nucleic acids or a mixture of the two?

I have already stated my reasons for preferring proteins. I prefer proteins partly because my model works well with ten species of monomer and works badly with four species, partly because amino-acids fit better than nucleotides the requirements of pre-biotic chemistry, and partly because I am attracted by the Margulis vision of parasitism as a driving force of early evolution and I like to put nucleic acids into the role of primeval parasites. None of these reasons is scientifically compelling.

At what stage did random genetic drift give way to natural selection?

The model has life originating by neutral evolution according to the ideas of Kimura (Kimura, 1970, 1983). A population crosses the saddle-point to the ordered state by random genetic drift. The model does not allow natural selection to operate because it does not allow the populations in cells to grow or to reproduce. So long as there is no birth and death of cells, there can be no natural selection. However, once a cell has reached the ordered state as defined in the model, it can go beyond the model and pass into a new phase of evolution by assimilating fresh monomers from its environment. A cell which increases its population N by assimilation will quickly become stabilized against reversion to the disordered state since the lifetime of the ordered state increases exponentially with N. It can then continue to grow until some physical disturbance causes it to divide. If it divides into two cells there is a good chance that both daughter populations contain a sufficient assortment of catalysts to remain in the ordered state. The processes of growth and division can continue until the cells begin to exhaust the supply of nutrient monomers. When the monomers are in short supply, some cells will lose their substance and die. From that point on, evolution will be driven by natural selection.

As soon as natural selection begins to operate, there will be an enormous advantage accruing to any cell which acquires the knack of dividing itself spontaneously instead of waiting for some natural process such as wave motion or turbulent flow to break it apart. At first, spontaneous division might be an accidental consequence of a tendency of the cell surface to weaken as the cell expands in volume. At a later stage, the weakening of the surface and the subsequent spontaneous division would become organized and integrated into the metabolic cycle of the cell. Cells would then be competing with one another in a straight-forward Darwinian fashion, with the prize of survival going to those which had learned to grow and divide most rapidly and reliably. In this way the processes of natural selection could have been well established long before the cells had acquired anything resembling the modern machinery of cell division.

Does the model contradict the Central Dogma of molecular biology?

The Central Dogma says that genetic information is carried only by nucleic acids and not by proteins. The dogma is true for all contemporary organisms, with the possible exception of the parasites responsible for scrapie and kuru. Whether or not the scrapie parasite turns out to be a true exception to the dogma, my model implies that the dogma was untrue for the earliest forms of life. According to the model, the first cells passed genetic information to their offspring in the form of enzymes which were probably proteins. There is no logical reason why a population of enzymes mutually catalysing each other's synthesis should not serve as a carrier of genetic information.

68

The question How much genetic information can be carried by a population of molecules without exact replication? is intimately bound up with the question of the nature of homeostasis. Homeostasis is the preservation of the chemical architecture of a population in spite of variations in local conditions and in the numbers of molecules of various kinds. Genetic information is carried in the architecture and not in the individual components. But we do not know how to define architecture or how to quantify homeostasis. Lacking a deep understanding of homeostasis, we have no way of calculating how many items of genetic information the homeostatic machinery of a cell may be able to preserve. It seems to be true, both in the world of cellular chemistry and in the world of ecology, that homeostatic mechanisms have a general tendency to become complicated rather than simple. Homeostasis seems to work better with an elaborate web of interlocking cycles than with a small number of cycles working separately. Why this is so we do not know. We are back again with the question Why is life so complicated? But the prevalence of highly complex homeostatic systems, whether we understand the reasons for it or not, is a fact. This fact is evidence that large amounts of genetic information can in principle be expressed in the architecture of molecular populations without nucleic acid software and without apparatus for exact replication.

How did nucleic acids originate?

I remark earlier (Chapter 2) on the curious fact that nucleic acids are chemical cousins of the ATP molecule, which is the chief energy-carrier in the metabolism of modern cells. I like to use this fact to explain the origin of nucleic acids as a disease arising in some primitive cell from a surfeit of ATP. The Margulis picture of evolution converts the nucleic acids from their original status as indigestible by-products of ATP metabolism to disease agents, from disease agents to parasites, from parasites to symbionts and finally from symbionts to fully integrated organs of the cell.

How did the modern genetic apparatus evolve?

The modern genetic apparatus is enormously fine-tuned and must have evolved over a long period of time from simpler beginnings. Perhaps some clues to its earlier history will be found when the structure of the modern ribosome is explored and understood in detail. The following sequence of steps is a possible pathway to the modern genetic apparatus, beginning with a cell which has RNA established as a self-reproducing cellular parasite but not yet performing a genetic function for the cell. (a) Non-specific binding of RNA to free amino-acids, activating them for easier polymerization; (b) specific binding of RNA to catalytic sites to give them structural precision; (c) RNA bound to amino-acids becomes transfer RNA; (d) RNA bound to catalytic sites becomes ribosomal RNA; (e) catalytic sites evolve from special-purpose to general-purpose by using transfer RNA instead of amino-acids for recognition; (f) recognition unit splits off from ribosomal RNA and becomes messenger RNA; (g) ribosomal structure becomes unique as the genetic code takes over the function of recognition. This is only one of many possible pathways that might have led to the evolution of the genetic code. The essential point is that all such pathways appear to be long and tortuous. In my opinion, both the metabolic machinery of proteins and the parasitic self-replication of nucleic acids must have been in place before the evolution of the elaborate translation apparatus linking the two systems could begin.

How late was the latest common ancestor of all living species?

The universality of the genetic code suggests that the latest common ancestor of all living creatures already possessed a complete genetic apparatus of the modern type. The geological record tells us that cells existed very early, as long as 3 eons ago. It is generally assumed that the earliest cells which are preserved as microfossils already possessed a modern genetic apparatus, but this assumption is not based on concrete evidence. It is possible that the evolution of the

70

modern genetic apparatus took eons to complete. The ancient microfossils may date from a time before there were genes and ribosomes. The pace of evolution may have accelerated after the genetic code was established, allowing the development from ancestral procaryote to eucaryotic cells and multicellular organisms to be completed in less time than it took to go from primitive cell to ancestral procaryote. It is therefore possible that the latest common ancestor came late in the history of life, perhaps as late as half-way from the beginning.

Does there exist a chemical realization of my model, for example a population of a few thousand amino-acids forming an association of polypeptides which can catalyse each other's synthesis with 80 per cent accuracy? Can such an association of molecules be confined in a droplet and supplied with energy and raw materials in such a way as to maintain itself in a stable homeostatic equilibrium?

These are the crucial questions which only experiment can answer.

What will happen to my little toy model when the problem of the origin of life is finally solved?

This is the last question raised by the model and it is easily answered. The answer was given nearly two hundred years ago by my favourite poet, William Blake (A Vision of the Last Judgment, Rossetti MS, 1810):

> To be an Error and to be Cast out
> is a part of God's design.

Wider implications

At the end of his book *What is Life?* Schrödinger put a four-page epilogue with the title 'On determinism and free will'. He there states his personal philosophical viewpoint, his reconciliation between his objective understanding of the physical machinery of life and his subjective experience of free will. He writes with a clarity and economy of language which have rarely been equalled. I will not try to compete with Schrödinger in summing up in four pages the fruits of a lifetime of philosophical reflection. Instead I will use my last pages to discuss some of the wider implications of our thoughts about the origin of life, not for personal philosophy but for other areas of science. I use the word science here in a broad sense, including social as well as natural sciences. The sciences that I have particularly in mind are ecology, economics and cultural history. In all these areas we are confronting the same question that is at the root of the problem of understanding the origin of life: Why is Life so Complicated? It may be that each of these areas has something to learn from the others.

The concept of homeostasis can be transferred without difficulty from a molecular context to ecological, economic and cultural contexts. In each area we have the unexplained fact that complicated homeostatic mechanisms are more prevalent, and seem to be more effective, than simple ones. This is most spectacularly true in the domain of ecology, where a typical stable community, for example a few acres of woodland or a few square feet of grassland, comprises thousands of diverse species with highly specialized and interdependent functions. But a similar phenomenon is visible in economic life and in cultural evolution. The open market economy and the culturally open society, notwithstanding all their failures and deficiencies, seem to possess a robustness which centrally planned economies and culturally closed societies lack. The homeostasis provided by unified five-year economic plans and by unified political control of culture does not lead to a greater stability of economies and cultures. On the contrary, the simple homeostatic mechanisms of central control have generally proved more brittle and less able to cope with historical shocks than the complex homeostatic mechanisms of the open market and the uncensored press.

But I did not come to Cambridge to give a political pep-talk in defense of free enterprise. My purpose in mentioning the analogies between cellular and social homeostasis was not to draw a political moral from biology but rather to draw a biological moral from ecology and social history. Fortunately, I can claim the highest scientific authority for drawing the moral in this direction. It is well known to historians of science that Charles Darwin was strongly influenced in his working-out of the theory of evolution by his readings of the political economists from Adam Smith to Malthus and McCullough. Darwin himself said of his theory: 'This is the doctrine of Malthus applied to the whole animal and vegetable kingdom.' What I am proposing is to apply in the same spirit the doctrines of modern ecology to the molecular processes within a primitive cell. In our present state of ignorance we have a choice between two contrasting images to represent our view of the possible structure of a creature newly emerged at the first threshold of life. One image is the hypercycle model of Eigen, that is to say a molecular structure tightly linked and centrally controlled, replicating itself with considerable precision and achieving homeostasis by strict adherence to a rigid pattern. The other image is the 'tangled bank' of Darwin, an image which Darwin put at the end of his *Origin of Species* to make vivid his answer to the question 'What is Life?', an image of grasses and flowers and bees and butterflies growing in tangled profusion without any discernible pattern, achieving homeostasis by means of a web of interdependences too complicated for us to unravel. The tangled bank is the image that I have in mind when I try to imagine what a primeval cell would look like. I imagine a collection of molecular species that are tangled and interlocking like the plants and insects in Darwin's microcosm. This was the image which led me to think of error-tolerance as the primary requirement for a model of a molecular population taking its first faltering steps toward life. Error-tolerance is the hall-mark of natural ecological communities, of free market economies and of open societies. I believe it must have been a primary quality of life from the very beginning. But replication and error-tolerance are naturally antagonistic principles. That is why I like to exclude replication from the beginnings of life, to imagine the first cells as error-tolerant tangles of non-replicating molecules, and to introduce replication as an alien parasitic intrusion at a later stage. Only after the alien intruder has been tamed is the reconciliation between replication and error-tolerance achieved in a higher synthesis, through the evolution of the genetic code and the modern apparatus of ribosomes and chromosomes.

The modern synthesis reconciles replication with error-tolerance by establishing the division of labour between hardware and software, between the genetic apparatus and the gene. In the modern cell, the hardware of the genetic apparatus is rigidly controlled and error-intolerant. The hardware must be error-intolerant in order to maintain the accuracy of replication. But the error-tolerance which I like to believe was inherent in life from its earliest beginnings has not been lost. The burden of error-tolerance has merely been transferred to the software. In the modern cell, with the infrastructure of hardware firmly in place and subject to a strict regime of quality-control, the software is free to wander, to make mistakes and occasionally to be creative. The transfer of architectural design from hardware to software allowed the molecular architects to work with a freedom and creativity which their ancestors before the transfer could never have approached.

The analogies between the genetic evolution of biological species and the cultural evolution of human societies have been brilliantly explored by Richard Dawkins in his book *The Selfish Gene* (Dawkins, 1976). The book is mainly concerned with biological evolution. The cultural analogies are pursued only in the last chapter. Dawkins' main theme is the tyranny which the rigid demands of the replication apparatus have imposed upon all biological species throughout evolutionary history. Every species is the prisoner of its genes and is compelled to develop and to behave in such a way as to maximize their chances of survival. Only the genes are free to experiment with new patterns of behaviour. Individual organisms must do what their genes dictate. This tyranny of the genes has lasted for three thousand million years and has been precariously overthrown only in the last hundred thousand years by a single species, *Homo sapiens*. We have overthrown the tyranny by inventing symbolic

74

language and culture. Our behaviour patterns are now to a great extent culturally rather than genetically determined. We can choose to keep a defective gene in circulation because our culture tells us not to let hemophiliac children die. We have stolen back from our genes the freedom to make choices and to make mistakes.

In his last chapter Dawkins describes a new tyrant which has arisen within human culture to take the place of the old. The new tyrant is the 'meme', the cultural analogue of the gene. A meme is a behavioural pattern which replicates itself by cultural transfer from individual to individual instead of by biological inheritance. Examples of memes are religious beliefs, linguistic idioms, fashions in art and science, in food and clothes. Almost all the phenomena of evolutionary genetics and speciation have their analogues in cultural history, with the meme taking over the functions of the gene. The meme is a self-replicating unit of behaviour, like the gene. The meme and the gene are equally selfish. The history of human culture shows us to be as subject to the tyranny of our memes as other species are to the tyranny of genes. But Dawkins ends his discussion with a call for liberation. Our capacity for foresight gives us the power to transcend our memes just as our culture gave us the power to transcend our genes. We, he says, alone on earth, can rebel against the tyranny of the selfish replicators.

Dawkins' vision of the human situation as a Promethean struggle against the tyranny of the replicators contains important elements of truth. We are indeed rebels by nature, and his vision explains many aspects of our culture that would otherwise be mysterious. But his account leaves out half the story. He describes the history of life as the history of replication. Like Eigen, he believes that the beginning of life was a self-replicating molecule. Throughout his history, the replicators are in control. In the beginning, he says, was simplicity. The point of view that I am expounding in these lectures is precisely the opposite. In the beginning, I am saying, was complexity. The essence of life from the beginning was homeostasis based on a complicated web of molecular structures. Life by its very nature is resistant to simplification,

75

whether on the level of single cells or ecological systems or human societies. Life could tolerate a precisely replicating molecular apparatus only by incorporating it into a translation system that allowed the complexity of the molecular web to be expressed in the form of software. After the transfer of complication from hardware to software, life continued to be a complicated interlocking web in which the replicators were only one component. The replicators were never as firmly in control as Dawkins imagined. In my version the history of life is counterpoint music, a two-part invention with two voices, the voice of the replicators attempting to impose their selfish purposes upon the whole network, and the voice of homeostasis tending to maximize diversity of structure and flexibility of function. The tyranny of the replicators was always mitigated by the more ancient cooperative structure of homeostasis that was inherent in every organism. The rule of the genes was like the government of the old Hapsburg Empire: 'Despotismus gemildert mit Schlamperei', 'Despotism tempered by sloppiness'.

One of the most interesting developments in modern genetics is the discovery of 'junk DNA', a substantial component of our cellular inheritance which appears to have no biological function. Junk DNA is DNA which does us no good and no harm, merely taking a free ride in our cells and taking advantage of our efficient replicative apparatus. The prevalence of junk DNA is a striking example of the sloppiness which life has always embodied in one form or another. It is easy to find in human culture the analogue of junk DNA. Junk culture is replicated together with memes, just as junk DNA is replicated together with genes. Junk culture is the rubbish of civilization, television commercials and astrology and juke-boxes and political propaganda. Tolerance of junk is one of life's most essential characteristics. In every sphere of life, whether cultural, economic, ecological or cellular, the systems which survive best are those which are not too fine-tuned to carry a large load of junk. And so, I believe, it must have been at the beginning. I would be surprised if the first living cell were not at least 25 per cent junk.

That is the end of my story, and it brings me back to the beginning. I have been trying to imagine a framework for the origin of life, guided by a personal philosophy which considers the primal characteristics of life to be homeostasis rather than replication, diversity rather than uniformity, the flexibility of the genome rather than the tyranny of the gene, the error-tolerance of the whole rather than the precision of the parts. The framework that I have found is an abstract mathematical model which is far too simple to be true. But the model incorporates in a crude fashion the qualitative features of life that I consider essential, looseness of structure and tolerance of errors. The model fits into an overall view of life and evolution that is more relaxed than the traditional view. The new and looser picture of evolution is strongly supported by recent experimental discoveries in the molecular biology of eucaryotic cells. Edward Wilson, who was also my illustrious predecessor as Tarner Lecturer in Cambridge (Wilson, 1982), describes the new picture of the eucaryotic genome as 'a rainforest with many niches occupied by a whole range of elements, all parts of which are in a dynamic state of change'. My philosophical bias leads me to believe that Wilson's picture describes not only the eucaryotic genome but the evolution of life all the way back to the beginning. I hold the creativity of quasi-random complicated structures to be a more important driving force of evolution than the Darwinian competition of replicating monads. But philosophy is nothing but empty words if it is not capable of being tested by experiment. If my remarks have any value, it is only insofar as they suggest new experiments. I leave it now to the experimenters to see whether they can condense some solid facts out of this philosophical hot air.

Bibliography

Anderson, P. W. (1983). Suggested model for prebiotic evolution: The use of chaos. *Proc. Nat. Acad. Sci. USA* **80**, 3386–90.

Cairns-Smith, A. G. (1982). *Genetic Takeover and the Mineral Origins of Life*. New York: Cambridge University Press.

Dawkins, R. (1976). *The Selfish Gene*. New York: Oxford University Press.

Dyson, F. J. (1982). A model for the origin of life. *J. Mol. Evol.* **18**, 344–50.

Eigen, M., Gardiner, W., Schuster, P. & Winckler-Oswatitch, R. (1981). The origin of genetic information. *Sci. Am.* **244**, No. 4, 88–118.

Kimura, M. (1970). Stochastic processes in population genetics. In *Mathematical Topics in Population Genetics*, ed. K. I. Kojima, pp. 178–209. Berlin: Springer.

Kimura, M. (1983). *The Neutral Theory of Molecular Evolution*. New York: Cambridge University Press.

Margulis, L. (1970). *Origin of Eucaryotic Cells*. New Haven: Yale University Press.

Margulis, L. (1981). *Symbiosis in Cell Evolution*. San Francisco: Freeman and Co.

Miller, S. M. & Orgel, L. E. (1974). *The Origins of Life on the Earth*. Englewood Cliffs, NJ: Prentice-Hall, Inc.

Niesert, U., Harnasch, D. & Bresch, C. (1981). Origin of life between Scylla and Charybdis. *J. Mol. Evol.* **17**, 348–53.

Oparin, A. I. (1957). *The Origin of Life on the Earth*, 3rd edn, translated by Ann Synge. Edinburgh: Oliver and Boyd.

Prusiner, S. B. (1982). Novel proteinaceous infectious particles cause scrapie. *Science* **216**, 136–44.

Schrödinger, E. (1944). *What is Life? The Physical Aspect of the Living Cell*. Cambridge University Press.

Spiegelman, S. (1967). An in vitro analysis of a replicating molecule. *American Scientist* **55**, 3–68.

Terzaghi, E. A., Wilkins, A. S. & Penny, D. (1984). *Molecular Evolution, an Annotated Reader*. Boston: Jones and Bartlett.

Von Neumann, J. (1948). The General and Logical Theory of Automata. Lecture given 1948. In *Cerebral Mechanisms in*

Behavior – The Hixon Symposium, ed. L. A. Jeffress, pp. 1–41. New York: John Wiley, 1951; and in *J. Von Neumann, Collected Works*, ed. A. H. Taub. Vol. 5, pp. 288–328. New York: MacMillan, 1961–63.

Wilson, E. O. (1982). Remarks quoted by R. Lewin. *Science* **216**, 1091–2.

Author index

Anderson, P. W., 33–4
Bresch, C., 38, 65
Cairns-Smith, A. G., 30, 32, 33, 34, 35
Dawkins, R., 74, 75, 76
Dyson, F. J., 44
Eigen, M., 3, 4, 5, 9–11, 12, 14, 18, 30, 31, 33, 34, 35, 36, 37, 38, 39, 40, 60, 62, 63, 64, 65, 73, 75
Gardiner, W., 10, 31, 62
Harnasch, D., 38, 65
Kimura, M., 16–18, 33, 44, 45, 46, 67
Margulis, L., 11–16, 18, 25, 67, 69
Miller, S. M., 15, 21, 22, 23

Niesert, U., 38, 39, 65, 66
Oparin, A. I., 30–1, 33, 34, 35, 37, 39, 40, 41, 44, 45, 51, 64
Orgel, L. E., 3, 4, 9–11, 12, 18, 21, 22, 23, 31
Penny, D., 25
Prusiner, S. B., 27
Schrödinger, E., 1, 2, 3, 4, 5, 6, 7, 9, 18, 65, 71, 72
Schuster, P., 10, 31, 62
Spiegelman, S., 62, 63, 64
Terzaghi, E. A., 25
Von Neumann, J., 6, 7, 9, 18
Wilkins, A. S., 25
Wilson, E. O., 77
Winckler-Oswatitch, R., 10, 31, 62